Writing Style and Standards
in Undergraduate Reports

Second Edition

by Sheldon Jeter
and Jeffrey Donnell

College Publishing books are printed on acid-free paper.

ISBN: 978-1-932780-06-2
Library of Congress Control Number: 2011901573

College Publishing
12309 Lynwood Drive, Glen Allen, Virginia 23059
Phone (800) 827–0723 or (804) 364–8410
Fax (804) 364–8408
Email collegepub@mindspring.com

Writing Style and Standards in Undergraduate Reports

Second Edition

by Sheldon Jeter
and Jeffrey Donnell

College Publishing

Glen Allen, Virginia

For Frank K. Webb

(Georgia Tech BSME, 1938)

ABOUT THIS GUIDE

This guide addresses two primary questions of inexperienced technical writers: "How should my report be written?" and "How do I present my experimental work appropriately using figures, tables, equations, spreadsheets, and the like?" The first two sections of this guide respond to these two questions.

The first section of this guide, *Writing Style in Undergraduate Reports*, speaks to the problems that students face as they draft their reports. Specifically, it outlines the norms of format in engineering reports, and it describes the way format is linked to the substance of the report. This section of the guide also offers a concrete review of the norms of paragraph and sentence formation as they apply to engineering reports. Style in Undergraduate Reports also provides a set of guides and example reports based on the writing requirements of the different laboratory courses at The Woodruff School of Mechanical Engineering of the Georgia Institute of Technology. These model reports are offered as examples of good reports, and they should be consulted when questions arise concerning format, style, or presentation of data.

The second section of this guide, *Standards for Undergraduate Reports*, speaks to the problems students face as they present their work using figures, tables, equations, and so forth. This section of the guide explicitly outlines the norms for assembling and labeling figural and tabular information, it provides examples of well-made figures and tables, and it provides detailed checklists to help students to determine whether their data is professionally presented.

The third section of this guide, *Writing on the Job*, speaks to the kinds of tasks students face when they make the transition from classroom reporting to workplace communication, where problems are often open-ended and audiences cannot be assumed to be engineering professionals. This section of the guide concentrates on a design report and presentation for an open-ended project. This report is prepared to describe a technical project to non-technical readers, and this section highlights the visual and verbal steps the author has taken to accommodate these non-technical readers.

TABLE OF CONTENTS

PART ONE
Writing Style in Undergraduate Reports

PART TWO
Standards for Undergraduate Reports

PART THREE
Writing on the Job

Part One:
Writing Style in Undergraduate Reports

Jeffrey Donnell

CHAPTER 1.1

GLOBAL STYLE: AN INTRODUCTION TO UNDERGRADUATE REPORTS

Abstract

The goal of this guide is to outline the standards of format and style that govern experimental engineering reports. These standards should be adequate for use in undergraduate laboratory courses and in typical professional reporting practice. Because the norms of format and style described here are well accepted in the industry, students should expect the material presented here to help them in their professional careers as well as in their undergraduate laboratory classes.

This guide addresses four areas of concern among beginning technical writers: judgment, format, clarity, and graphics. First, to help students make good judgments about their reporting responsibilities and about their audience's needs, in Chapter 1.1, the introduction describes the relationship between experimentation and reporting, and it outlines some of the stylistic hallmarks of good technical communication. Second, in order to help students manage the organization and appearance of their documents, this guide's "Global Style" section gives a detailed outline of the norms of report formatting and of the information that is appropriately presented under each format heading. Third, so as to help students write clearly and directly, this guide's sections on "Paragraph Style" and "Sentence Style" outline the basics of low-level text management. Finally, to help students introduce figures and tables that are professional in appearance and easy to comprehend, this guide's "Attachments" section gives explicit directions for the preparation, labeling, and presentation of figures tables and calculations.

Introduction

This introductory section addresses the judgments students must make as they begin to develop technical reports concerning experimental projects. Specifically, it indicates how experimental engineering pertains to technical communication, it explains how technical writing should support the experimental substance of an experimental report, it outlines the kinds of substantive matters that your audience expects you to discuss, and it outlines the stylistic norms that all technical documents must honor.

What is technical writing?

Good technical writing is orderly, and it is predictable. Well written technical documents follow a well-established order (also called *format*), and they predictably present information according to established conventions. When a student masters the basics of format and the conventions of presenting information, that student will be able to present even the most complicated material in technical reports that are easy to prepare and easy to comprehend.

The goals of technical reports

People write different kinds of documents to meet different kinds of goals. In business communication, for example, reports and analyses are sometimes used as sales tools. For example, the goal of a stock analyst's report might be to help brokers sell shares of a company to clients. The client might read the report in order to locate information that might justify a purchase of shares, and the report might be designed in order to highlight that information. Technical reports are prepared under different circumstances in order to meet different goals. Technical reports document experimental investigations that solve problems or answer questions. The specific goal of the technical report is to describe explicitly and briefly what specific problem(s) or question(s) were addressed, what data was collected, how that data was collected, how that data was analyzed, and how it was used to draw a conclusion.

How does experimental engineering pertain to technical reporting?

When you conduct an experimental investigation, you are expected to view your data objectively, to collect and evaluate that data accurately, and to develop conclusions that are specific and concrete. When you report on any investigation, you must describe procedures, analytical methods, and results using words that are at once objective, specific, and concrete. These qualities are valued by the professionals who will one day review, evaluate, and reproduce your work. Objectivity, specificity, and concreteness can be difficult to define and teach, but we know that these qualities are always found in reports that are orderly in presentation, logical in discussion, and impersonal in mode.

Making reports orderly

Reports are orderly when they are systematically organized; in technical reports, organization is linked to document format, which will be outlined in detail in Part II of this guide. It is sufficient here to say that good technical reports adhere to formatting norms because those norms reflect the processes of orderly investigations. In undergraduate investigations, technical report format is generally governed by the scope of the project and the method of inquiry used on the project; consequently, the report format will exactly reflect the steps of thinking—of experimentation and of observation—that governed the investigation. Since investigations follow well-established methods of observation and data collection, reports are prepared following a well-established format.

Making reports logical

When you adhere carefully to the norms of technical report format, your readers will find that your reports are easy to read because they present information in a predictable fashion. However, the author of a report must do more than simply use prescribed format headings in a prescribed sequence; the author of a report should also speak to the logic that underpins the report's format and that has driven the investigation itself. To establish this logic, report authors need to speak briefly to the connections between the elements of information contained in the different sections of their reports. Usually, such connections can be established by inserting into each section of the report a brief response to the question: "How will this information be used elsewhere in this report?" Readers perceive reports to be logical only to the extent that they can find explicit answers to this question in each section of the report.

Making reports impersonal

Engineering reports are project-oriented and impersonal. An impersonal report is not about the author(s), and it is seldom about people. Rather, an engineering report provides an issue-by-issue description of a project. In experimental engineering, our investigations answer questions about systems, about materials, and about the models people have developed to describe systems, materials, and so forth. The reports prepared to document such investigations must present the answers to those questions and the steps taken to obtain those answers. Technical reports discuss the individuals participating in the investigation only in the rarest of circumstances.

Experimental Engineering and Communication Style

Experimental engineering projects are conducted in two stages: first, experiments are conducted to solve problems; then, a report is written to describe how the experiment was conducted. Both experimentation and writing are organized around these questions:

> What was the objective (or problem)?
> How was the investigation performed?
> What data was collected?
> What result was achieved?
> What conclusion was reached?

These questions are fundamental to engineering. These questions must be answered in order to plan and conduct an investigation; in turn, the engineers who read reports expect to find descriptions of how the investigation was planned and conducted. When the report provides complete and detailed answers to these questions, other engineers are assured that all contingencies have been checked and that the report's conclusions are reliable. Furthermore, a complete report gives instructions to any future engineer who might seek to reproduce the work. All professionals must provide such instruction to everybody who might review or reproduce the work.

Conventions for Presenting Technical Substance

Any presentation of data requires the author to do two things well. First, the author must describe the data systematically, and second, the author must prepare text that makes use of well-known conventions in the profession. The next two sections of this Guide offer a general outline for presenting your data in a systematic fashion that makes use of professional conventions.

Presenting data

All reports present data collected during experiments, and they provide text descriptions of that data as well. Both the experimental description and the presentation of data need to satisfy technical readers that the data are objective and reliable. To that end, any description of procedures and apparatus should indicate that the instrumentation was properly selected and used. Additionally, the report must indicate that raw data was not accepted at face value; a renowned astrophysicist (Eddington 1936) has rightly stated that "We [should] not believe our eyes unless we are at first convinced that what they appear to tell us is credible." Your presentation of data must include an explanation of the point the data illustrates, and it should include a physical argument to justify that explanation.

Discussing data

Data should be discussed thoroughly, and the discussion should acknowledge any anomalies present in the data. If a figure shows, for example, a negative efficiency, the discussion of that figure should comment on this oddity. If the data shows a divergence with expected results or theory, this divergence should be acknowledged in the text. Because of time constraints, lab procedures cannot always be repeated, so sometimes you must report anomalous data. In such situations, you should address and explain the problems, and you ought to propose logical explanations for them.

Describing data

When data is presented, the report must make a point or draw a conclusion about that data. A good experimental finding is a detailed, numerical comparison between empirical data and a theoretical or hypothetical model. This comparison needs to be substantive. Unacceptably superficial observations look like this: "The empirical profile diverged from the theoretical model." A more appropriate comparison describes the confirming or opposing trends in detail: "The empirical profile of normalized velocity is nearly uniform with minor decrease *of about 5% at .9 normalized radius*. In contrast, the theoretical velocity ratio varies continuously and substantially, *decreasing to .4 at .9 normalized radius*."

Evaluating data

Results are useful only to the extent that they can be explained. Consequently, in addition to grounding descriptions in numbers, authors need to explain on a scientific basis, why their data agree or disagree with the proposed hypothetical or theoretical model. Authors need to provide valid physical reasons for agreement, and these reasons must be supported by pertinent physical evidence. Many students try to explain away discrepancies between experimental and predicted results through the use of the phrase "experimental error." Other students rely on purely mathematical explanations when their direct observations are somehow unacceptable. Both kinds of argument are weak and will be ignored by any sophisticated reader.

Precision

Claims should be substantiated with numbers when they are available. When presenting findings or conclusions, you need to avoid unsupported assertions or conjectures, and you must avoid vague or unquantified statements. The best way to stay out of trouble is to rely on numbers instead of adjectives and adverbs. For example, avoid statements like this: "The data agree *reasonably well* with the model." Instead, say this: "The data agreed with the model *within +/– 10%*."

Accuracy

Numbers must be accurate, but they must be reasonable as well. For almost any calculation, computing tools make it possible to quote long strings of trailing digits. If you quote a velocity, for example, of "2.141492654 m/sec," you demonstrate only that you have access to lots of number-crunching capacity. Numbers are not necessarily accurate or useful simply because they are long, so follow the rules for identifying significant digits that are presented in Part II and report only significant digits in text and tables.

Text Conventions of Technical Reports

Audiences read technical reports in order to locate information, and they expect this information to be packaged in accordance with professional conventions. These conventions involve not only the organization of the report, but also the way points are articulated, the way procedures and results are described, and the way the author manages tenses and sentence structures. Your best guides to these conventions will be those exemplary documents you encounter in your field, and those experienced colleagues and supervisors who can guide you while you gain experience. Until you gain such experience, you can rely on the detailed suggestions that follow later in this guide, and on the practical guidelines that follow.

Technical writing is impersonal

Lab reports and project reports should speak objectively about problems, experiments, and results. The people who obtain these results should not appear as the topics of the reports or as the subjects of long strings of sentences in those reports. The following sample paragraph italicizes first-person references to show how a project team violates this convention:

> *The team* was asked to determine the RMS and peak-to-peak voltages. *We* were uncertain at first how to measure these voltages, but after discussion with *our teaching assistant, we decided* to measure these voltages directly from the oscilloscope screen. *We decided* to conduct this part of the experiment with the frequency set to 1 kHz on the signal generator, and *we decided* to set the CD component to 3 volts. After many adjustments and false starts, *we measured* a peak voltage of 0.84 volts. *Then we* found the RMS value to be 0.29 volt.

The following sample paragraph presents the same information impersonally by removing the personal references. The verb forms characteristic of this impersonal stance are italicized:

> The RMS and peak-to-peak voltages *were directly measured* on the oscilloscope, using a frequency of 1kHz on the signal generator, with the DC component of the signal set to 3 volts and the AC component oscillating at 1kHz. The peak-to-peak value *was found to be* 0.84 volts and the RMS value *was found to be* 0.29 volt.

Technical writing is neither wholly active nor wholly passive

The grammatical terms *active* and *passive* describe verb placement as well as verb tense. In active constructions, the verb is placed near the beginning of the sentence, following a short subject structure. Engineers use active constructions regularly: "The pressure increases as the temperature rises."

In passive constructions, the verb is placed near the end of the sentence and a sentence subject is often dropped. Passive constructions are commonly used in order to keep reports impersonal. Your lab notes, for example, might show the following sentence: "I recorded the air pressure, temperature, and humidity." Rendered in passive construction, this sentence becomes an impersonal statement: "The air pressure, temperature, and humidity were recorded."

Technical writing uses both present and past tenses

Reports describing experimental projects are generally written after the project has concluded. Such reports typically rely on past tense verbs because past tense accurately reflects the relationship between the time of report presentation and the time of the investigation. As do all forms of communication, engineering reports use past tense to describe events that have been completed and that are particular.

However, engineering reports use present tense verbs quite often. Present tense verbs are used to describe events that are ongoing and to describe concepts (or principles or laws) that are considered to be universally true. Consequently, when an experiment verifies the accuracy of some theoretical model, the report will use present tense verbs to show that this model is still considered accurate: "The results obtained in this investigation indicate that the model accurately predicts the response of this system."

On occasion a mathematical proof is offered as part of a project report. Proofs are also presented in present tense, both because they present information that is considered to be universally true, and because the proof itself is structured as a set of instructions to be followed by the reader during the process of reading.

Technical writing concludes and it explains

When a technical question has been posed, it must be answered and discussed in the text. For example, if you are asked whether a fluid flow is laminar or turbulent, you must do more than to cite the Reynolds number. You must describe the shape of the velocity profile, the distribution of turbulence intensity, and any other observations. Conclusions must be based on experimental evidence, and reports should indicate that experimental observations confirm (or dispute) a theoretical prediction, such as exceeding the critical Reynolds number for transition to turbulence.

Reports must be complete

All reports have at least four sections, and all of these sections must be completed in each report. Readers expect to see an introductory problem statement, a description of methods, a separate description of results, and a closing summary. They want to see this material in full, even if they already know the problem and the conclusion ahead of time. Readers expect to see complete reports even if some of the material is not new to them.

Avoid usage problems

Technical readers prefer to read text that is unambiguous and direct. You can maintain these qualities in your text by observing these suggestions concerning usage.

1. Avoid using "this" as a pronoun. When you start a sentence with "This," your reader needs to see the answer to the question: "This What?"

2. Avoid using parentheses except for documentation citations. Parentheses mark clauses that the author wants but does not need. The reader wants nothing that the author does not need.

3. Learn how to use the subjunctive. The subjunctive mood is characterized by a plural, past tense verb used with a singular subject. It is used to mark hypothetical statements. Avoid statements like this: "If the temperature *was* lower, the result *might be* different." Write this instead: "If the temperature *were* lower, the result *would be* different. Indeed, the temperature was lower during the second section of the project . . ." The subjunctive marks hypothetical statements and if-then statements such as these: "If the temperature were lower. . . ." This statement might be followed by an active, past tense sentence like this: "Indeed, the temperature was lower during the second run."

Reports matter

Good results are of no value until they are communicated to colleagues and superiors. Technical reports should not be created at the last minute; they are as much a part of the project work as the acquisition of sample materials and the calibration of machines.

Global Style at the Report Level

Report format is governed by the logic of the investigation. Reports are built around the same set of questions and answers as are projects; reports raise and answer these questions in roughly the same sequence as they should arise during the project. Consequently, the report author's job is simply to articulate the questions or issues that were addressed in work, to articulate the responses that were developed for these questions, and to explain how conclusions were reached.

The format summary below outlines the format sections used in most technical reports. For each of these format sections, the general questions that a professional should answer in that section are listed, and a few detailed prompts are provided in order to help the student translate general requirements into specific chunks of text.

Format headings for technical reports

Technical reports are most often divided into subsections using the following headings:

1. Objectives
2. Apparatus and Procedures
3. Data (also called Findings)
4. Analysis
5. Discussion
6. Closure
7. Sample Calculations (not always a separate section)

Each subsection of a report describes a portion of the project, such as the experimental procedures, the results, and the analysis that substantiates your conclusions. In fact, each subsection responds to a particular question or set of questions about the project, including "What were you trying to do?" "How did you collect data?" and so forth. This material will seem obvious and trivial to most students, as it will have been the subject of a recent lecture or class handout. However, it is still your job to present such "obvious" information. Your experimental reports describe your project; they do not ask you to determine whether you did something "new and interesting." The specific questions posed by each subsection are detailed below, along with some possible responses from you:

Objectives

The Objectives section answers questions pertaining to the project's goals. The primary questions are these: 1. "What was the overall goal of the project?" and/or 2. "What values or relationships were to be determined?" The answers to these questions will usually involve "obtaining measurements" or "determining relationships"; in laboratory classes, the answers will often begin with a summary of the specific requirements listed in the class handouts.

Apparatus and Procedures

The Apparatus and Procedures section responds to questions concerning the methods of conducting the investigation: 1. "How did you collect the data that helped you make your determinations?" and/or 2. "What did you measure and how did you measure it?" Depending on the scope of the project, the answers to these questions might be very short, or they might be very long. Research projects often require lengthy discussions of procedures and equipment; reports describing routine quality assurance tests can usually describe methods in a sentence or two.

Data

The Data section should present a separate results description for each of the requirements, or goals. Each such description should respond to two questions, in order: 1. "What was done to collect data?" and 2. "What result, or data was obtained?" Typically, the first question is answered with descriptions of the materials used, or the settings on test or experimental equipment, and with a listing of any inputs or stresses introduced to the system under study. Responses to the second question typically involve introduction and discussion of an exhibit such as a figure or a table.

Analysis

The Analysis section should describe how data was used to reach a conclusion. Thus, the Analysis section presents responses to these questions: 1. "What equations or models were used?" 2. "What experimental values were used for which variables in these equations?" 3. "What final conclusion is drawn based on this analysis?"

In practical terms, if your objective is to determine the density of an object, your Data section should contain the dimensions and the mass (or weight). In your Analysis section, you would use this information to calculate volume and then density. The Analysis section should contain any theoretical calculations such as Simulink models, but it should not look like a scratch pad. The Analysis section should explain how you set up your calculations, but it should also contain a reference to the Sample Calculations section, where you will show your work more fully.

From time to time the Analysis section is omitted, particularly if a project involves observation only. On other occasions, the Analysis section may be merged with the Discussion section. When this happens, you should still plan to respond to the three questions listed above; you should simply view the third question as the opening or introduction to the Discussion.

Discussion

The Discussion section explicitly states and supports any conclusions that can be drawn from the material shown in the Data and the Analysis sections. Specifically, the Discussion section should present your responses to questions like these: 1. "Do the results compare well with expected or predicted values? 2. If not, what might account for these discrepancies?" In responding to such questions, it is appropriate for authors to comment on problems making measurements and to indicate what such problems mean with regard to the objective of the project.

Closure

The Closure section is often loosely called a *conclusion*, but it should be treated as a brief summary of results. In this section of a report, authors should summarize any important findings, results, or conclusions that they have presented earlier in the report. No new information or speculation should be introduced in the Closure section.

Sample Calculations

When showing calculations, students must do the following things. First, they must show the equations to be used. Second, they must show the steps in finding the solution. Third, they must show the final result. To do this correctly, students must first write the basic equation in terms of variables; then, they must rewrite that equation substituting experimental values with units for variables. Finally, they should show the calculated answer with units, paying attention to significant figures.

After the numerical calculation has been shown, the student must describe in detail how the equations were used, which experimental values were used for each variable, and what physical forces or structures were represented by each variable or expression in the equation. Shown below is a sample calculation pertaining to an Oscilloscope investigation. Please notice that this example devotes three lines to equations, it devotes two text lines to explanations at the end, and it devotes a further three lines to text descriptions of what physical entity is represented by each variable in the equation:

SAMPLE CALCULATIONS

Let the input signal be described by

$$V(t) = A \sin(\omega t) \tag{1}$$

where $A = V_{pp}/2$ is the signal amplitude, V_{pp} is the peak-to-peak voltage amplitude, ω is the angular frequency, and $f = 1/T$, where f is the frequency and T is the period of the signal. The root-mean-square of the signal is defined as:

$$V_{rms} = \left[\frac{1}{T} \int_0^T V^2(t)\, dt \right]^{1/2} = \left[\frac{1}{T} \int_0^T A^2 \sin^2(\omega t)\, dt \right]^{1/2}$$
$$= \left[\frac{A^2}{T} \int_0^T \frac{1 - \cos(2\omega t)}{2}\, dt \right]^{1/2} = \left[A^2/T \left(\frac{T}{2} - 0 \right) \right]^{1/2} = \left[\frac{A^2}{2} \right]^{1/2} = \frac{A}{\sqrt{2}} \tag{2}$$

Note that this result, $V_{rms} = A/\sqrt{2}$, is valid only if the input voltage $V(t)$ is sinusoidal. For instance, for a square wave, it is easy to show that $V_{rms} = A = V_{pp}/2$.

Format and Information in Common Report Types

The most common kinds of reports and supporting documents are:

- Abstracts
- Summaries (also called "executive summaries")
- Itemized Reports
- Narrative Reports

While these documents have different names, they all require responses to the formatting questions listed in the previous section of this guide. These report forms differ mainly in the amount of detail the author is expected to supply in response to these questions, although the level of detail expected from an author is also linked to the scope and complexity of the project. For example, students will write itemized reports for introductory courses, where they are asked to perform a number of procedures or calculations whose answers are either well-known or are relatively easy to determine. In industry, a professional might prepare an itemized report to describe a set of routine quality assurance tests on a particular system or piece of equipment.

Narrative reports are prepared when projects are larger, when they are more fully focused on a particular technical issue, or when the author has more responsibility, as in upper-level experimental courses. In a capstone course, on the other hand, students are asked to work independently on a relatively complicated project. In such a circumstance, students will prepare a long narrative report, which is both lengthy and complex; professionals can expect to develop such longer reports when they begin to do independent research projects. Other classes and other projects will fall in between these poles, according to the complexity of the project and amount of responsibility given to the students.

The remainder of Part One of this guide presents samples of abstracts, summaries, itemized reports, and narrative reports, with short, format-driven commentaries.

Abstracts and Executive Summaries

Abstracts and executive summaries are short documents placed at the very beginning of reports. While they are almost always attached to report packages, they should be treated as separate documents, often placed on a separate sheet from the introduction sections of the reports. Abstracts and Summaries are designed to give busy readers clues about the content of your reports. They do not, however, provide specific references to exhibits or to particular pages in the report.

Abstracts

Abstracts are very short; they are usually about one-third page long, and they are results- and conclusion-oriented. The abstract primarily describes—in order—the problem or goal, the method, and the result. It does this in perhaps four or five sentences, and those four or five sentences generally describe the contents of the main format sections of your report: The goal or problem statement, the result, the methodology, and the analysis. In short, your abstract should provide one-sentence answers to these questions:

1. What problem was addressed?
2. What result was obtained?
3. How was the study conducted?
4. How was data evaluated?

The sample abstract below was prepared for a capstone project report, and it demonstrates all of the features required in an appropriate report abstract:

SAMPLE ABSTRACT

This project determined the thermal characteristics of a high density polyethylene and of a particular type of styrofoam, two materials commonly used in manufacture of drinking cups. Since the goal of such cups is to insulate the contents, the thermal properties are important factors in cup design.

The thermal conductivity of the styrofoam was found to be 0.414 W/mK. The thermal conductivity of the high density polyethylene was found to be 0.448 W/mK.

These values were obtained by experimentally measuring the temperature of a liquid held in each cup as the cups were exposed to a hot water bath. These experimental values were then used to calculate thermal conductivity.

Summaries

Summaries are long versions of abstracts. A summary differs from an abstract in that a summary provides a more detailed problem statement (often with motivation), it provides a longer treatment of methods, and it can give action recommendations, including suggestions for future research, estimates of budget or of production schedules, and so forth. Summaries commonly describe the contents of each of the main sections of a report; a summary may devote only a sentence or two to each section of the report, and it will often include exhibits, such as figures or tables. Consequently, summaries are stand-alone documents, and many readers will treat them like very small reports. That is to say, readers read a summary when they want to be able to understand your work and your results by reading the summary but not the report itself.

In your undergraduate classes, most report summaries will be between three-quarters of a page and one full page in length, and they will resemble itemized reports, as described later. A one-page summary typically has three or four paragraphs, and these paragraphs respond to the four questions discussed earlier in the following way:

Paragraph 1:	What task was assigned?
Paragraph 2:	What steps were performed to address task #1?
	What result was reached on this task?
Paragraph 3:	What steps were performed to address task #2?
	What result was reached on this task?
Paragraph 4:	What conclusion was reached?

The sample summary below describes an investigation performed in an intermediate experimentation course.

SAMPLE SUMMARY

The goal of this project was to develop a very stable fluid flow system for demonstration of laminar flow in an undergraduate fluid dynamics laboratory. An existing fluid flow system was modified in order to meet two particular objectives:

1. The modified system must maintain stable turbulent flow and stable laminar flow through a circular glass tube,
2. After modification, the system must be characterized by use of a Laser Doppler Velocimeter (LDV).

Both objectives were met at the completion of this project.

Analysis of the existing system showed that flow variations were caused by the close integration of the pump, which was directly connected to the glass measurement pipe. This direct connection caused pulsating flows and allowed mechanical vibrations to be transmitted from the pump to the measurement tube. The existing control valve may also have contributed to turbulence in the fluid flow. To address these problems the pump was isolated from the system by introduction of a constant head tank, and a variable flow meter with a modulating valve was installed.

Flow profiles for the modified system were measured using an LDV. Using the flowmeter to adjust fluid velocity, flows at nine different average velocities were characterized, ranging from 0.0092 m/s to 0.64 m/s. The overall result was a set of laminar and turbulent profiles for Reynolds Numbers ranging from 380 to 26400.

The results demonstrate that the modification was successful. The modified system produced Reynolds Numbers as low as 400, it produced stable turbulent flows at Reynolds Numbers up to 23000, and it produced stable laminar flow for Reynolds Numbers as high as 1300. For Reynolds Numbers above 1300 the system exhibited transitional behavior of the type usually expected at Reynolds Numbers around 2100. To address this problem of premature transition, it is recommended that the head tank be enlarged, that a diffuser be added to the supply line, and that the supply line for the flow into the head tank be moved farther from the outlet.

Preparing Itemized Reports

When projects are short and are governed by a very specific list of small tasks, the results are usually presented in the form of the so-called itemized report. In industry, itemized reports are developed to describe procedures that are both repetitive and common, such as qualification or acceptance tests. A professional might, for example, be asked to test a product or system according to the ASME Performance Test Code. This test code actually gives general guidelines, which are then implemented with a specific set of measurements and calculations. These tests are performed in the field, then a report is developed, providing a very brief description of the different tests. The itemized report format is ideal for describing the results obtained during tutorial experiments in lab classes, and for describing the results obtained during routine testing.

Itemized reports should attend carefully to the questions associated with the format standard followed in this guide:

1. Objectives
2. Apparatus and Procedures
3. Data (sometimes called Findings)
4. Analysis
5. Discussion
6. Closure
7. Sample Calculations (not always a separate section)

Itemized reports are preferred in introductory experimentation courses, and they are used for prelab assignments from time to time in other courses. A sample itemized report from an introductory experimentation course is included in a separate section of this guide. In that report, each format item is broken into a separate heading, making the presentation somewhat long. An experienced author, however, will be able to compress the goals, procedures, results, and analysis into a single paragraph of a few sentences. When this happens, the format and information content of the report will be unchanged; only the amount of space devoted to each item will vary.

Preparing Narrative Reports

Narrative reports tell the whole story of a project. A narrative report actually may be prepared using either of two formats: the letter report and the technical memorandum. No matter which term is used, such reports describe small but focused experimental projects. Short reports usually are longer than itemized reports, in part because they usually describe more complicated projects. A short report might be used to describe an investigation responding to questions like this: "How can this system be modeled?" or "Does this model hold under these conditions?" In response to such questions, the report describes a project whose several tasks work together to characterize a system or to test a model. The report makes a sustained case for the conclusions reached over the course of the project.

Short reports typically describe small projects (or project tasks) that were developed under supervision. If a student happens to work on a faculty member's research project, for example, that student might be given a general goal or a question to answer, and then be left to design and conduct an experiment that meets that goal or answers that question. The short report form is used to describe projects in an intermediate experimentation course.

Depending on the scope of the project, a short report's description of experimental methods and of data evaluation might run to some length. A brief closing section will restate conclusions.

The format for a short report is essentially the same as that for an itemized report. These two report forms differ mainly in that each main format section of the short report will provide more detailed discussion than will an itemized report. A section of an itemized report may be only a sentence or two in length, while the corresponding section of a short report may run to a page or more. Beyond this, the short report follows our standard format list.

Objectives
1. What was the overall goal of the project?
2. What values or relationships were to be determined?

Apparatus and Procedures
1. What instruments were used to collect data?
2. What was the method for collecting data?

Data (also called Findings)
1. What was done to collect data? (only for details not described under Procedures)
2. What result, or data was obtained?

Analysis

1. What equations or models were used to evaluate data?
2. What experimental values were used for variables in these equations?
3. What conclusion is drawn based on this analysis?

Discussion

1. How do the results compare with expected or projected values?
2. How are discrepancies between real and predicted values explained?

Closure

1. What important findings are mentioned in the Analysis and/or Discussion sections?

Sample Calculations (not always a separate section)

About Long Reports

Long reports are a variety of narrative report, differing in scale more than in organization. Such reports may be called technical reports when published in a journal, or they may be called project reports when they are intended for limited distribution.

People prepare long reports to describe large projects that have been conducted independently. In such situations, the author of the report is usually the principle investigator on the project, and the report's readers generally are not completely familiar with the project's technical details. Consequently, long reports are common on research projects, even if a supervising faculty member originally assigned the project. As research supervisors, faculty members often find it easier to read complete reports, and from time to time they pass these reports along to colleagues who are less familiar with the particulars of the project. A project report is required as the final submission in most capstone courses, as it presents the results of a student team working over an entire term.

Long reports are appropriate under either of two conditions. First, long reports are appropriate for large and complex projects, for the author may have many procedures to describe, and the author must also explain how all the parts of the project fit together. Second, long reports are necessary for most readers who have not been involved in the daily work of the project. If a student completes a project during one afternoon in a well-supervised lab, then that student probably will not write a long report. If, however, the project was spread across most of a term, then a long report is probably expected. A useful schematic for format and content in Long Reports follows:

Objectives

1. What was the overall goal of the project?
2. What values or relationships were to be determined?

Apparatus and Procedures

1. What instruments were used to collect data?
2. What was the method for collecting data?

Data (also called **Findings**)

1. What was done to collect data? (only for details not described under Procedures)
2. What result, or data was obtained?

Analysis

1. What equations or models were used to evaluate data?
2. What experimental values were used for variables in these equations?
3. What conclusion is drawn based on this analysis?

Discussion

1. How do the results compare with expected or projected values?
2. How are discrepancies between real and predicted values explained?

Closure

1. What important findings are mentioned in the Analysis and/or Discussion sections?

Sample Calculations (not always a separate section)

Style in Paragraphs

Paragraphs are crucial building blocks in technical reports because they contain the identifiable blocks of useful information. In presenting this information, a well-written paragraph also focuses on and develops a topic. If your paragraphs develop topics effectively, they will be easy to follow and understand. However, if your paragraphs do not attend to topic development, your reports will resemble grocery lists; they may be accurate, but they will not be easy to read.

In itemized reports and in many short reports, the business of whole format sections may be compressed into a single paragraph. That is to say, a single paragraph may need to name a task or problem, present a method and a result, and then make some point about that result. To complicate matters, each such paragraph must appear to be internally cohesive, and it must somehow maintain the *flow* of the overall report.

In order to manage these complicated problems, you should treat each paragraph as if it were a mini technical report with a format as fully structured as you use for a whole technical report. This paragraph format is driven by questions that pertain to the topic developed in the paragraph. This paragraph format has three major components, one of which is optional. These paragraph format components are listed below:

1. Issue
2. Discussion
3. Closure

Each of these paragraph elements responds to some question or set of questions concerning the topic of the paragraph. The specific questions are outlined in the following three sections.

Paragraph issues

The issue section of a paragraph corresponds roughly to the Goals and Procedures section of a technical report, for it responds to the question: "What problem or task is addressed?" In long reports describing complex projects, a paragraph issue section might also indicate what section of the project the problem pertains to. In so doing, it responds to the question: "How does this problem fit into the overall project?" The paragraph issue differs from the report introduction, of course, because a given paragraph should address only one of the constituent problems or tasks addressed by the investigation.

In addition to naming a problem or task, the paragraph issue section makes some sort of claim or point about the problem; it answers the questions: "What result is obtained?" and/or "What was learned?" In undergraduate lab investigations, this point is usually a description of the result obtained for the task under discussion.

Paragraph discussions

The discussion section of a paragraph either explains or supports the point offered in the paragraph issue. That is to say, the paragraph discussion section responds to one or more of the following questions:

1. How was the result obtained?
2. How is this claim substantiated?
3. Why was this decision made? (or why was this conclusion drawn?)
4. How might this open question be addressed?

The discussion section is by far the longest section of a paragraph, for it contains the details that respond to these questions. Depending on the nature of the project, responses to these questions might direct the discussion in any of the following directions:

1. Description of an experimental procedure
2. Description of data analysis techniques
3. Description of experimental constraints and/or analytical logic
4. Description of work to be done in the future

Paragraph closure

Paragraph closure sections are optional. When used, they are inserted in order to explain the relationship between a given paragraph and some subsequent paragraph or section of the report. Such closure sections are seldom needed in itemized reports, where tasks are usually described in full with a single paragraph. However, in long reports describing projects with many constituent tasks, it may be necessary to indicate how a given set of results or procedures are to be used in a subsequent section of the report.

Topic development

A paragraph tells a story about a chunk of a project, a piece of equipment, for example, the management of data, or the procedure for collecting data. In order to communicate its story, the paragraph must be perceived by the reader to be focused on some topic or issue, and it must be perceived to develop some story about that topic or issue. Readers perceive a paragraph to be focused only when they *see* the paragraph use the topic word or name at the heads of sentences. The sample paragraph below successfully describes a test to determine the real behavior of a set of thermocouples used in an experiment. This paragraph can be said to be cohesive, to flow, and to show logical development *only* to the extent that its sentences begin repeatedly with terms that are prominent in previous sentences. When readers see such repetition, they consider that the paragraph tells a story about the terms that are repeatedly placed at the beginnings of these sentences.

SAMPLE PARAGRAPH

The goal of this first experimental step was to characterize the behavior of the thermocouples to be used in the rest of the project. Specifically, readings obtained from the three thermocouples that will later be placed in the test samples were compared to the readings from the thermocouple that will be placed in the water bath alone and from the thermocouple that will measure the skin temperature of the container. The three sample thermocouples were arranged in approximately the positions they will take in a spherical sample of the experimental material. They were then placed into the bath and allowed to settle to a constant temperature. The thermocouples detected slightly different readings for the bath temperature. These different readings are shown in Figure 6, which also indicates that the bath temperature thermocouple recorded a temperature of fully one degree Celsius below the reading recorded from the thermocouple that recorded skin temperature. Further, the skin temperature thermocouple showed a temperature one degree Celsius below both the sample midpoint and the sample center thermocouples. Figure 9 displays these results graphically. These results indicate that the thermocouples must be calibrated to agree at a common temperature in order for subsequent sample tests to provide precise results.

Paragraph Structure
A Quick Reference Sheet

(**Main components are in bold.** Underpinning questions are in roman type.)

Issue (Issues announce topics, raise questions, and make claims.)
1. What project task or question is addressed in this paragraph?
2. What claim is made (or What result is presented)?

Discussion (Discussions describe tasks, explain answers, or substantiate claims.)
1. How is the result obtained?
2. How is the claim substantiated?
3. Why was this decision made? (or Why was this conclusion drawn?)
4. How might this open question be addressed?

Style in Sentences

Technical reports should be written with language that is simple, direct, and descriptive. While technical reports address concepts that sometimes seem hard to grasp, these concepts can and should be described using terms that are short, simple, and can be written without internal punctuation. Even the variables that we use in complicated equations have names that pertain to physical entities or forces, and these names are typically short, simple, and can be written with a small number of keystrokes. Your reports need to be written in such a way that they are easy to read even for an engineer who has not participated in your project. In order to write such reports, you need to pay attention to the way you assemble every sentence.

In technical reports, sentences need to be impersonal, empirical, and objective. These sentence qualities are directly linked to the subjects, verbs, and modifiers from which all sentences are constructed. The following paragraphs discuss these qualities in detail.

Impersonal structures

Impersonal structures are formed from sentence subjects. In the subject position of each sentence, you answer the question "Who or what is this sentence about?" Members of the research team (that's you) rarely appear in sentences describing their projects. The subjects of your sentences should be the things (or forces) you are studying.

Avoid this: "We found that the pressure varied with changes in temperature."
Do this: "Pressure varied with changes in temperature."

Empirical structures

Empirical structures are formed with verbs. In the verb position of each sentence, you answer the questions "What happened?" or "What did [the subject/agent] do?" The answers commonly involve active verbs—things *deflect, break, boil, cool,* and so forth.

21

Experiments require observation, so you should expect to use some verbs associated with seeing. Phenomena are *observed, seen, found,* and *shown,* for example, and values are *calculated* and *determined.* These are things that you do. But because you do not get to appear as subjects in your sentences, you must sometimes use the so-called passive verb constructions in order to maintain an impersonal stance while recounting your work faithfully. Passive constructions appear occasionally in impersonal texts.

Avoid this: "We used Equation 4 to determine the Reynolds number."
Do this: "Equation 4 was used to determine the Reynolds number."
Or: "The Reynolds number was determined using Equation 4."

Objective structures

Objectivity in sentences is maintained with modifiers. The best known modifiers are adjectives and adverbs. With these structures we characterize the things we observe and the changes we observe; modifiers help us to answer the "How" questions that drive science reporting: "How large?" "How many?" "How fast?" In response to these questions, the reader needs to see numbers.

Avoid this: "A large sample was heated to a high temperature for a long time; then it was cooled quickly."
Do this: "A 70 gram sample was heated to 200 degrees C for 4 hours. It was then cooled to 0 degrees C over the course of 20 minutes."

Managing verb tenses

Many students are confused about the relationship between tense management and passive constructions. We can get past this confusion by remembering one principle: past tense constructions are different from passive sentence constructions. Past tense constructions are used to describe any action taken in the past. When a report describes an experimental investigation, it will be written in past tense because the experiment was performed in the past.

The so-called passive construction is primarily a sentence inversion which places the sentence's *object* in front of the sentence's *subject.* When sentences are inverted in this form, the subject of the sentence is often dropped. Technical reports often use passive constructions in order to maintain an impersonal stance. A passive/inverted construction drives the example shown earlier in the section, Empirical structures:

Avoid this: "We used Equation 4 to determine the Reynolds number."
             This is first person and it is past tense.
Do this: "Equation 4 yielded the Reynolds number."
             This construction is impersonal, and past tense.
Or: "The Reynolds number was determined using Equation 4."
             This is still a past tense statement; because it is also inverted, it is also called passive.

22

Report Format
A Quick Reference Sheet

Main headings are in bold. Underpinning questions are in roman type.

Objectives

    1. What was the overall goal of the project?

    2. What values or relationships were to be determined?

Apparatus and Procedures

    3. What instruments were used to collect data?

    4. What was the method for collecting data?

Data (also called **Findings**)

    5. What was done to collect data? (Only for details not described under Prodecures)

    6. What result, or data was obtained?

Analysis

    7. What equations or models were used to evaluate data?

    8. What experimental values were used for variables in these equations?

    9. What conclusion is drawn based on this analysis?

Discussion

    10. How do the results compare with expected or projected values?

    11. How are discrepancies between real and predicted values explained?

Closure

    12. What important findings are mentioned in the Analysis and/or Discussion sections?

Sample Calculations (not always a separate section)

CHAPTER 1.2

GUIDE TO ITEMIZED REPORTS

Objectives and Procedures

The goal of this chapter is to provide a model of a well-written report for a project in an introductory experimentation course, to show how that model report is linked to the requirements and the report outline for a typical introductory project, and to explain how the figures and tables in the report make use of accepted standards for visual communication.

The Report

Lab reports for introductory courses are typically composed of numerous smaller reports. The discussion below will show first how to format one such report; then it will explain the standards for managing text and style; and finally it will describe how to label, insert, and cite figures and tables.

The text of the report is constructed of brief responses to the prelab requirements. If more than one response is required, the responses must be numbered, as are the paragraphs in this section of the guide. All exhibits, such as spreadsheets or figures, must be cited in the text. Exhibits must have unique, consecutive numbers, and they must have descriptive captions.

At least one numbered paragraph must be devoted to each requirement, even if the paragraph contains only a sentence or two naming the result and citing the pertinent exhibit. A paragraph is required because every exhibit must be cited somewhere in the text.

Format for Itemized Reports

Reports in introductory experimentation courses are commonly divided into sections using headings such as the following:

1. Objectives and Procedures
2. Results (or sometimes Findings)
3. Analysis
4. Discussion
5. Sample Calculations
6. References

Each subsection of a report provides a narrative description of the experimental procedures, the results (and their interpretation), and the analysis that substantiates your conclusions, and each subsection must show sample calculations (as needed). In fact, each subsection responds to a particular question or set of questions about the project, including "What were you trying to do?" "How did you collect data?" and so forth. This material will seem obvious to most students, as it will have been the subject of a recent lecture or class handout. However, it is still your job to describe such "obvious" information. Your experimental reports describe what you did; they do not ask you to determine whether you did something "new and interesting." The specific questions posed by each subsection are detailed in the following text, along with some possible responses from you.

Objectives and Procedures

The Objectives and Procedures section answers the questions: "What was the overall goal of the project?" and "What was I asked to determine?" The answers to these questions should always involve *obtaining measurements* or *determining relationships*. In fact, your answers should include the specific requirements listed in the project handouts; our model reports list and number these requirements.

The Objectives and Procedures section must also respond to the question: "How did you collect the data that helped you make your determinations?" Specifically, "What did you measure and how did you measure it?" An efficient and appropriate response to this question appears in the last sentence of the Objectives and Procedures section of the sample report: "The procedure consisted of measuring directly on a two-channel oscilloscope the input and output voltages of the amplifier as the input voltage was varied from –12 to +12 volts."

Results

The Results section should present a separate results description for each of the requirements or goals. These results descriptions should respond to two questions, in order: 1. "How was data collected?" and 2. "What result, or data was obtained?" The attached sample report shows that one should respond to the first question by describing the materials used, the settings on the equipment, and/or any inputs introduced to the system under study. This model report also shows that one responds to the second question by introducing an exhibit—a graph, a table, or another type of figural information.

Analysis

The Analysis section should explain how data was processed in order to provide a conclusion. Thus, the Analysis section asks for responses to these questions: 1. "What equations were used?" 2. "What experimental values (obtained from Results) were used for each of the variables in these equations?" and 3. "What final conclusion was reached?"

In practical terms, if the objective is to determine the density of an object, the Results section should contain the dimensions and the mass (or weight). The Analysis section should use this information to calculate volume and then density. The Analysis section should contain any theoretical calculations such as simulation models, but it should not look like a scratch pad.

The Analysis section should explain how calculations were set up, but it should also contain a reference to the Sample Calculations section, where you must show your work more fully.

From time to time, the Analysis section is omitted, particularly if a project involves observation only. On other occasions, the Analysis section may be merged with the Discussion section. When this happens, you should still plan to respond to the three questions listed above; you should simply view the third question as the opening or introduction to the Discussion.

Discussion

In the Discussion section, you explicitly state any conclusions that you are able to draw based on your Results and Analysis sections. Such conclusions should address questions such as these: "Do the results compare well with expected or predicted values? If not, how can discrepancies be explained?" In responding to such questions, it is appropriate to comment on problems that arise in making measurements and to indicate what such problems mean with regard to the objective of the project.

Sample Calculations

In presenting the Sample Calculations section, it is best to describe in detail the equations that were used, which experimentally obtained values were used for which variables, and what physical forces or structures are represented by each variable or expression in an equation. As you examine the sample calculation shown in the Oscilloscope project, please notice that we devote three lines of text to equations, we devote two lines of text to explanation at the end, and we devote three lines of text to an explanation of what each variable represents.

When you present your sample calculations, you must first write the basic equation in terms of variables. Then, you should rewrite that equation in terms of values with units. Present your answer with units, and pay attention to significant figures. With the instructor's permission, you may use neat handwriting; you should not spend a great amount of time typing complicated equations.

Text Norms
Page design

Lab reports are commonly prepared in parts by lab partners. However, the final submission must look like a single document prepared by a single author during a single writing session. To assure that partners' contributions merge seamlessly, please observe the following requirements for page design and style:

- Use Times New Roman font, 12 point, 1.5 line spacing.
- Set margins for 1 inch on all sides of the page.
- Use ".doc" extensions for your files. (Avoid rich text format ".rtf.")
- Use boldface characters for section headings.
- Spellcheck reports and then proofread them.
- Check all of your numbers to make sure you have the correct number of significant digits.

Itemized reports should be prepared using normal paragraph style, which has two characteristics: 1. a single blank line is inserted to separate paragraphs, and 2. the first line of each paragraph is indented. In addition to these characteristics, you are asked to observe the general standards for titles, headings, and subheadings. Indented subheadings for paragraphs, called run-in side heads, should be underlined or bolded, they should be followed by a period, and they should be immediately followed by the text. Section headings should appear alone on a line, they should be bolded, and they should be aligned on the left margin. A heading should follow a blank line, and text should begin on the next line. Main titles should be bolded, centered, and set off from surrounding text by blank lines above and below.

Exhibits
Overview of exhibits
In the different subsections of your reports, results are presented or substantiated using exhibits. Exhibits are nontext pieces of information; they include—but are not limited to—figures, tables, sketches, and sometimes sample calculations (as discussed earlier). In the model report attached here, the exhibits are included, or integrated, in the body of the report. From time to time, however, students will be asked to place exhibits at the end of a report as attachments. When your exhibits are packaged as attachments, the rest of your report will not change, except for the obvious and trivial matters of spacing on the page. The rest of this section outlines the standards you must observe when you use the most common professional exhibits—figures and tables.

Figures
The figures in lab reports are typically graphical representations of data, as are Figures 1–5 in the sample "Lab Zero" report. Many people loosely speak of such figures as *graphs* or *charts*; such loose speaking is incorrect, however. Your reports will show your results in figures.

Every figure in your report must have a figure number and a descriptive caption. On rare occasions authors are allowed to save space by using a single caption for a number of closely related figures—1a, 1b, 1c, and so on. Always check with the instructor before doing so. The figure number and the caption are always placed on the same line, below the figure. The figure number is placed to the left of the caption. Students may sometimes be allowed to place a Title above the figure, but any use of such titles does *not* free the student of the requirement to place a figure number and caption below the figure.

The axes in figures must be labeled, and each label must show the units used on that axis. If a figure includes more than one data set, you should include a legend to indicate which data set is which. In all figures, scales and tick marks should be adjusted appropriately; if the data starts at zero, then the range of the graph should start at zero, not at –50.

Students should always cite the figure in the text before it is shown on the page, and they should always refer to a specific figure with a capital letter "F". Figures 4 and 5 in the sample report demonstrate explicitly how to make, label, and number figures using single data sets.

Each figure is cited, or named, in the last sentence before the figure is displayed on the page, using a statement such as this: "The results are shown below in Figure 1." Figures should be numbered in the order that they are introduced in the report; the first figure you cite should be "Figure 1," the second figure should be "Figure 2," and so forth. When figures are attached at the back of a report, they must still be numbered and arranged in order of use. An example of a figure and its preceding text citation is provided below.

Sample Figure

<u>DC response.</u> Based on the measured values of output voltage vs. input voltage shown in Table 1, a plot of the DC response of the amplifier is shown below in Figure 4.

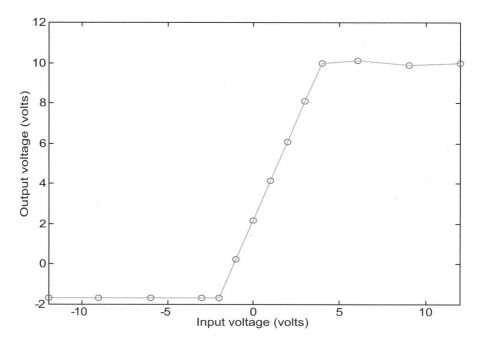

Figure 4. DC response of the amplifier.

Tables

Tables are to be treated with the same care and precision as are figures. The rows and columns in your tables must be clearly defined and labeled, and units must be visible either at the tops of the columns or at the left edges of the rows. Like figures, tables must have numbers and descriptive captions, with the table number placed to the left of the caption. Unlike figures, however, this number/caption line must be placed *above* the table. As is the case for figures, tables must be placed and numbered sequentially and in the order of use. Tables should be cited in the text *before* they are shown on the page, and specific tables must be referred to with a capital letter "T." Table 2 in the sample report demonstrates how to number, caption, and label tables; please notice that the last sentence before the table is introduced is designed only to cite the table, as shown by the statement that introduces the sample table.

SAMPLE TABLE

Table 2 below shows the measured input and output voltages with the calculated gain.

Table 2. Frequency response of the amplifier

Frequency	V_{in} (volts)	V_{out} (volts)	K	Gain (db)
10	1.0	5.0	5.0	14.0
10^2	1.0	9.8	9.8	19.8
10^3	1.0	10.1	10.1	20.1
10^4	1.0	10.0	10.0	20.0
5×10^4	1.0	10.1	10.1	20.1
8×10^4	1.0	10.6	10.6	20.5
9×10^4	1.0	10.7	10.7	20.6
10^5	1.0	10.1	10.1	20.1
2×10^5	1.0	6.0	6.0	15.6
10^6	1.0	3.0	3.0	9.5

Spreadsheets

If a spreadsheet is required in the prelab report, include as an attachment a printout of all important parts of the preliminary spreadsheet. The spreadsheet should be as complete and tested as possible. The spreadsheet should have a unique number and a descriptive title; this information should be centered and prominent on the page, as shown here:

Attachment 1. Experimental Spreadsheet

It is best to paste the spreadsheet block to a report page as a picture. For instructions on this technique, refer to Part Two, "Standards," of this manual. The heading can then be a mere line in your report text, which will be easy to edit and update.

Usually, a prelab assignment requires you to illustrate an important calculation with a block of formulas. If so, append a small block of representative cell formulas on a separate page. Do not print a large block of formulas along with less relevant data, such as numerical data, labels, and so forth. Rather, print a pertinent, representative block implementing the critical calculation or algorithm. Always describe this calculation in an accompanying text paragraph.

References

If used, references must be cited in the text and listed in a labeled section at the end of the text. The standard format for references can be found in a later section of this guide. Formats for references can also be found in common handbooks, such as *The Chicago Manual of Style*.

Appendices

Use appendices or other attachments for long mathematical derivations or calculations. Such attachments must be machine printed.

Final editing

Do not rely exclusively on your spell check program; always proofread your reports. Review the editing guidelines summary checklist on report preparation presented in Chapter 2.5 before you submit your reports.

We will now turn our attention to a model of a complete lab assignment from an introductory experimentation course. The project is titled "Laboratory #0," and the model presents the specific project requirements at the beginning of Chapter 1.3. The model report then describes the action taken on these requirements. Chapter 1.4 then presents an example of a short itemized report.

CHAPTER 1.3

EXAMPLE OF A LONG ITEMIZED REPORT

"Laboratory 0"
Requirements for Oscilloscope Investigation

INTRODUCTION TO THE OSCILLOSCOPE

I. OSCILLOSCOPE

The experimental arrangement is shown in Figure 1. A function generator will be connected to an oscilloscope using a BNC terminated coaxial cable. Select a frequency of 1 kHz and observe the signal on the oscilloscope. Be sure an appropriate amplitude and time scale are selected. Switch between AC and DC coupling and sketch the corresponding signals. Change the amplitude and DC offset settings on the function generator and observe the effect on the oscilloscope. Measure the peak-to-peak voltage, V_{pp}, and the root mean square voltage, V_{rms}, for a particular setting of the function generator.

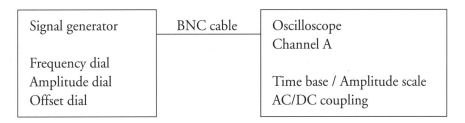

Figure 1. Experimental arrangement to operate the oscilloscope

II. CHARACTERIZING AMPLIFIER PERFORMANCE

The experimental arrangement shown in Figure 2 is used to operate the amplifier and observe some of its characteristics. A function generator provides a voltage signal to the amplifier, V_{in}. The signal output from the amplifier, V_{out}, is displayed on the oscilloscope.

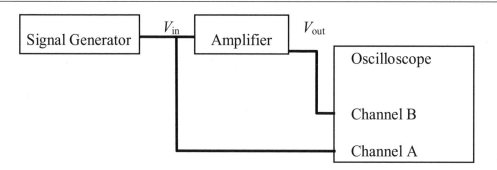

Figure 2. Experimental arrangement to operate the amplifier and observe its characteristics

A. DC operation (*i.e.*, when V_{in} and V_{out} are constant)

Using a table such as this, record the output voltage amplitude of the amplified signal, V_{out}, when sweeping the input voltage between −12 and +12 volts.

V_{in} (volts)					
V_{out} (volts)					

B. AC operations, *i.e.*, when the input voltage is oscillating at a given frequency

Set the function generator to "sine wave" with zero offset. Measure the frequency response of the amplifier between 10 Hz and 1 MHz on a log scale, *i.e.*, at 10, 100, 10^3, 10^4, 10^5, and 10^6 Hz by measuring the ratio $K = (V_{out}/V_{in})$, where V_{out} and V_{in} are peak-to-peak voltages. K is the amplification factor, and $G = 20 \log_{10} K$ is defined as the gain in decibels. Note that K and G may vary with frequency.

C. Clipping

Set the frequency to 1 kHz. Increase the input voltage and observe and sketch the resulting waveform, *i.e.*, the shape of the amplified signal.

Record your input values and your observations using a table such as this.

Frequency (Hz)	V_{in} (volts)	V_{out} (volts)	K	G (dB)

I. DYNAMIC ELECTRICAL MEASUREMENTS

A. Objectives and Procedures

The objectives of this experiment were to observe the effect of a direct current (DC) offset on an input signal, and to describe the effects of alternating current (AC) and DC coupling on the output of an oscilloscope.

To reveal the functions of the DC offset, AC coupling, and DC coupling, the voltage generated by a function generator was observed directly on a digital oscilloscope. Sketches were made of waveforms to record the effects of these functions on the output signal. Also, mean voltage, peak-to-peak voltage, RMS voltage, and frequency were measured with the oscilloscope to see the effects on these values.

B. Experimental Results

The DC offset control on the function generator was found to control the magnitude of the average, or mean, voltage. It was also noted that neither the peak-to-peak voltage nor the frequency varied with DC offset. The RMS voltage did vary, but this is expected, since it is related to the average absolute value of the voltage. Table I.1 displays the measured quantities for a sinusoidal signal without DC offset and sinusoidal signal with a 3-volt DC offset.

Table I.1 Comparison of measured values for sinusoidal signals with and without DC offset

	with	without
DC offset (V)	0.00	3.00
Mean voltage (V)	0.00	3.00
Peak-to-peak voltage (V)	0.84	0.84
Frequency (kHz)	1.0	1.0
RMS voltage (V)	0.29	3.02

Figure I.1 shows a sketch of a signal with zero DC offset, whereas Figure I.2 is a sketch of a signal with a 3-volt DC offset. Note that the vertical scales in the two figures are different.

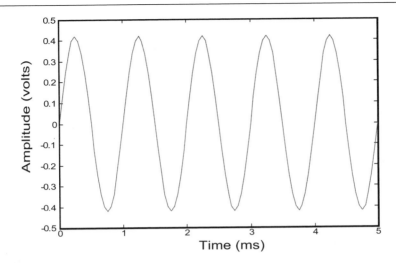

Figure I.1. Sketch of a sinusoidal (AC) voltage with no constant (DC) offset.

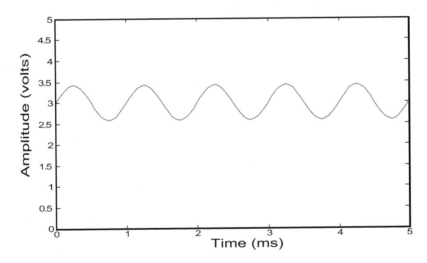

Figure I.2. Sketch of same sinusoidal (AC) voltage signal shown in
Figure I.1 after a 3-volt DC offset has been added.

The signal from Figure I.2 was continued, and the oscilloscope was set to DC coupling. Figure I.3 is a sketch of the waveform observed on the oscilloscope with DC coupling.

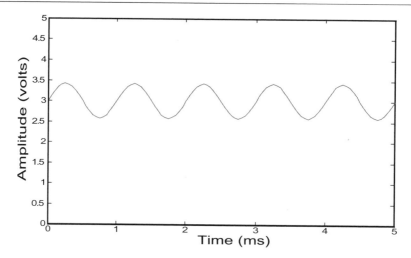

Figure I.3. A sketch of the waveform observed on the oscilloscope with DC coupling shows the sinusoidal (AC) voltage offset by a constant (DC) voltage.

Next, the input signal was held constant, but the oscilloscope setting was changed from DC to AC coupling. Although the input signal still had a 3 volt DC offset, the waveform on the oscilloscope screen was centered on 0 volts. Figure I.4 below is a sketch of the waveform observed on the oscilloscope with AC coupling.

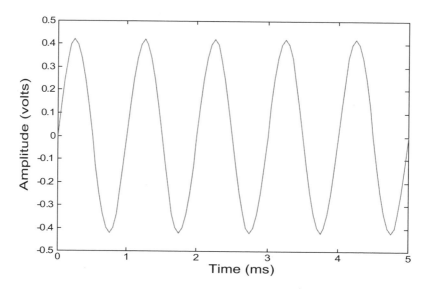

Figure I.4: A sketch of the waveform observed on the oscilloscope with AC coupling shows the sinusoidal (AC) signal with no apparent offset.

C. Discussion

The DC offset control on the function generator was found to control the magnitude of the mean voltage of the signal. No DC offset meant that the average value of the signal was zero. The positive magnitudes and the negative magnitudes of the signal were identical. The signal was "centered on" zero. By adding a DC offset of 3 volts, the signal was "moved up" to be "centered on" 3 volts.

Figure I.3 illustrates that DC coupling shows both the DC component and the AC component of an input signal. In contrast, Figure I.4 indicates that AC coupling shows only the AC component of the input signal. AC coupling eliminates the DC offset by using a capacitor or some other arrangement of circuitry. This may cause problems for very low frequency oscillations as they may be treated as approximately constant, or DC, voltage, depending on the time constant of the circuit.

In this experiment, voltage was measured two different ways. Peak-to-peak voltage is simply the difference between the maximum and minimum values of an oscillating signal. The root-mean-square voltage is more complicated. The name is derived from the mathematical calculation used to determine the RMS value.

Let the sinusoidal input signal with no DC offset, $V(t)$, be described by Equation (I.1),

$$V(t) = A\sin(\omega t) \tag{I.1}$$

where A is the signal amplitude, ω is the angular frequency, and t is time. The root-mean-square of the signal, V_{rms}, is defined by equation (I.2),

$$V_{rms} = \left[\frac{1}{T} \int_0^T V(t)^2\, dt \right]^{1/2} \tag{I.2}$$

where T is the period of the signal. For a specific time interval, the signal amplitude is squared at each moment. The mean of these squared values is evaluated using the integral. The square root of this mean is the RMS value.

Substituting Equation (I.1) into Equation (I.2) gives the relationship between the amplitude of a sinusoidal signal and its RMS value:

$$V_{rms} = \left[\frac{1}{T} \int_0^T V(t)^2\, dt \right]^{1/2} = \left[\frac{1}{T} \int_0^T A^2 \sin^2(\omega t)\, dt \right]^{1/2} = \frac{A}{\sqrt{2}} \tag{I.3}$$

The measured peak-to-peak voltage was 0.84 volts. The amplitude, A, is half the peak-to-peak voltage, V_{pp}, and

$$V_{rms} = \frac{V_{pp}}{2\sqrt{2}} = \frac{0.84 \text{ volt}}{2\sqrt{2}} = 0.30 \text{ volt} \tag{I.4}$$

The measured RMS value of 0.29 volts is within 3% of the above-calculated value. The difference could be due to inconsistencies in the input signal. Unfortunately, any fluctuations in the measured RMS value were not recorded during the experiment.

CHAPTER 1.4

EXAMPLE OF A SHORT ITEMIZED REPORT

TO: Instructor   9 April 2003
FROM: Felix A. Learner, ME 4053-B
SUBJECT: Report for Calorimeter Experiment

1. <u>Experimental Spreadsheet</u>. An EXCEL spreadsheet has been designed and implemented to receive and process the experimental data. The spreadsheet, including representative data, is appended as Attachment 1. A block of the pertinent data is reproduced below as Table 1.

Table 1. Heat Capacity Data and Model

temp	data	model
C	J/kg-K	J/kg-K
10.1	2005	2000
20.0	2030	2041
29.9	2105	2100
40.2	2180	2182
50.0	2285	2278
59.9	2390	2394

2. <u>Regression Modeling</u>. A regression model of the form,

$$C_p = \left(\frac{dh}{dT} \right)_P = C_0 + B_1 T + B_2 T^2 \qquad (1)$$

has been developed for the representative data. The model and the data are illustrated in Figure 1.

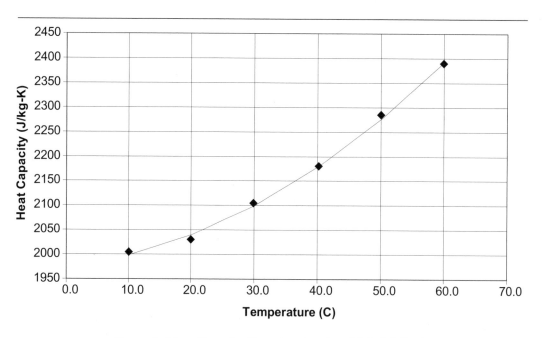

Figure 1. Heat Capacity Data (markers) and Model (line)

Attachment 1. Example Experimental Spreadsheet

File: HEATCAP		29 Sep 97, SMJ				
Example spreadsheet illustrating data processing, regression, and presentation.						
Summary Results:						
	alpha risk =	0.007				
	R squared =	0.998				
Example Data:					1	2
	temp	data	model		tc	tc^2
	C	J/kg-K	J/kg-K			
	10.1	2005	2000		10.1	102.0
	20.0	2030	2041		20.0	400.0
	29.9	2105	2100		29.9	894.0
	40.2	2180	2182		40.2	1616.0
	50.0	2285	2278		50.0	2500.0
	59.9	2390	2394		59.9	3588.0

SUMMARY OUTPUT

Regression Statistics	
Multiple R	0.998976397
R Square	0.997953841
Adjusted R Square	0.996589735
Standard Error	8.766268858
Observations	6

ANOVA

	df	SS	MS	F	Significance F
Regression	2	112440.2909	56220.14546	731.5809576	9.25569E-05
Residual	3	230.5424091	76.84746969		
Total	5	112670.8333			

	Coefficients	Standard Error	t Stat	P-value	Lower 95%	Upper 95%
Intercept	1976.921714	15.82745332	124.9045992	1.13145E-06	1926.551646	2027.291781
X Variable 1	1.303830113	1.03686987	1.257467451	0.297565826	-1.99595567	4.603615896
X Variable 2	0.09445367	0.014508043	6.510434915	0.007360909	0.048282559	0.140624782

CHAPTER 1.5

GUIDE TO NARRATIVE REPORTS

Introduction

Many undergraduate and professional projects require brief reports in the narrative format, examples of which are found in Chapters 1.6, 1.7, and 1.8. Narrative reports describe projects that are larger and more complex than are the highly scripted projects described by itemized reports. The distinguishing mark of a report in memorandum format is that it replaces the standard cover sheet with a memorandum heading or with a block title that may be formatted to resemble letterhead. In either case, this memorandum heading or title block must identify the project, the course and section, the group members, and the date the report was completed. Every contributor to the report should initial the heading to indicate that he or she has reviewed and approved the contents of the report.

Short narrative reports often accompany oral presentations. In such situations, the short report may be viewed as a kind of abstract whose job is to notify readers of the information that is to be found in the accompanying oral presentation. When used as abstracts in this way, short narrative reports are particularly useful for managers and supervisors who find themselves listening to progress reports on several projects a day; the extended abstract helps a busy manager to stay oriented and focused with a minimum of effort.

When you prepare a narrative report for use in the role of an abstract, you should be aware of these general expectations for size and substance:

- Such reports can be as long as two pages; they should not be longer than two pages.
- The report text can cite information that is to be presented orally.
- The report can include figures in the text.
- The report can have attachments.
- The report should have the same headings and subheadings as does the oral report.

Organization

The typical narrative report is organized into several brief sections, which may be as short as a single paragraph. In classes, each section of the report typically addresses a topic specified in the lab manual. Section headings are required. A typical list of section headings follows, each type being a brief characterization of the expectations that attend each section.

The Report
Heading
The heading of a narrative report should provide the same information as does the cover sheet to a written report. Specifically, the heading should show the following:

- The title of the project
- The names of the investigator(s)
- The course and section numbers
- The date

Introduction
The introduction describes the scope of the project by responding to the question "What were you trying to do?" In response to this question, you should identify the experiment and define its overall goals and constituent tasks. The introduction should also specify the place and date of the experiment. Introductions commonly restate information that is placed in report abstracts.

Overall observations, such as general trends and specific results, may be briefly summarized at the end of an introduction section. Such summaries are common and helpful in long reports that describe large projects. Even so, they must also be brief; you should strive to compress introductory material into no more than two sentences.

Apparatus and Procedures
This section responds to the question "How was data collected?" In response to this question, you should describe generically the equipment you used and the steps you went through during the project. In order to keep this document short, you should describe your methods only in the most general terms, mentioning the ranges of data that you collected—without going into further detail.

Collection and Presentation of Data
The body of the report responds to questions such as these: "What data was collected?" "How was it evaluated?" "What conclusion can be drawn?" You should respond to these questions systematically for each of the assigned tasks in your investigation; a complete set of responses to questions, plus supporting attachments, will form a section of the report. A section may be a single paragraph or a group of closely related paragraphs. Over the course of your report, you must develop sections to characterize your work on all experimental tasks, and you should cite and discuss all pertinent exhibits, attachments, and references.

Closure
This final section should offer a concise review of the project's tasks, followed by a concise statement of the important physical findings obtained for each task. Quantitative detail is not required at this point in a report, but a characterization of important trends, relationships, and confirmations is appropriate. This final section of a report does not introduce new information.

References

References must be cited in the text and listed in a labeled section at the end of the text. Guidelines for citing and documenting references are provided in Chapter 2.6.

Production and Editing

Page and paragraph formatting, exhibits, and production qualities are important elements of professional presentation. In addition to the general guidelines presented earlier, consider the following guidelines on format, exhibits, attachments, and editing when you prepare your reports.

Regular paragraph style

Reports for your classes should be prepared using regular paragraph style, which has two distinguishing characteristics: a single blank line is inserted to separate paragraphs, and the first line of each paragraph is indented. In addition to these characteristics, you should observe the general standards for titles, headings, and subheadings. Indented subheadings for paragraphs, called run-in side heads, should be underlined or bolded; they should be followed by a period; and they should be immediately followed by the text. Headings should appear alone on a line, they should be bolded, and they should be aligned on the left margin. A heading should follow a blank line, and text should begin on the next line. Main titles should be bolded, centered and set off from surrounding text by blank lines above and below.

Attachments and Exhibits

The specifics of your work will be described in figures, tables, spreadsheets, pictures, and the like. When you attach such material to a report, or when you insert it into the text, then it becomes an exhibit. Exhibits must have numbers and descriptive captions, and they must be inserted or attached in the order in which you direct the reader's attention to them. Spreadsheets use a prominent, centered header as a descriptive caption. Table numbers and captions are placed above the tables, while figures are numbered and captioned below the figure.

Spreadsheets

If a spreadsheet is required in the prelab report, include as an attachment a printout of all important parts of the preliminary spreadsheet. The spreadsheet should be as complete and tested as possible. The spreadsheet should have a unique number and name (or caption); this information should be centered and prominent on the page as shown here:

Attachment 1. Experimental Spreadsheet

The header utility can be used to generate this title, or the spreadsheet can be inserted as a picture element with the number and title added above it as text.

It is preferred that exhibits be incorporated into the text of your reports. When an exhibit such as a spreadsheet cannot be imported into your report, you may attach it at the back of the report.

Appendices

Use appendices or other attachments for long mathematical derivations or calculations. Attach appendices after the text.

Length

Long narrative reports should not exceed four printed pages, exclusive of figures and tables; short narrative reports should fill no more than two pages, exclusive of figures and tables. In all cases, you should not waste space and time on insignificant issues, and you should not discuss routine details of procedures.

Final editing

Carefully proofread your report. Do not rely exclusively on your spell checker.

CHAPTER **1.6**

EXAMPLE OF A SHORT NARRATIVE REPORT IN MEMORANDUM FORMAT

TO: Undergraduate Lab Instructional Staff   9 April 2003
FROM: Ernest L. Scholar, Section A
SUBJECT: Experiment on the Period of a Pendulum

Objective

The objective of this experiment was to measure the period of a simple pendulum experimentally and to compare the obtained data with a theoretical estimate of the period. The experiment was conducted in the instructional lab on 8 January 2003.

Procedure and Apparatus

Procedure

The procedure was to first measure the dimensions and mass of the pendulum to allow an estimate of its period. Then the pendulum was then placed on a knife edge and put in motion. The period was measured by manual timing for four repeated 50-cycle tests. The results and analysis of these tests are given below.

Apparatus

The pendulum is a flat steel bar with a drilled hole for the pivot, which is a knife edge. In this experiment the particular pendulum identified by Serial No. 3 was used. The dimensions of the pendulum were measured with a vernier caliper, specifically a Starrett model No. 123 with an uncertainty of about .05 mm (.002 inch), and a measuring tape, specifically a True Value type MMS425 with an uncertainty of about 1 mm (1/32 inch). The period was measured with a quartz crystal wrist chronometer, specifically a Timex Indiglo. The fractional uncertainty of the chronometer is about 6.0×10^{-6} (15 seconds/month, Timex, 1996).

Measurements

Dimensional Measurements

The dimensions of the pendulum were measured as follows. The overall length was measured to be 349 mm (13.8 in.) with the measuring tape. The thickness was measured to be 13.1 mm (0.514 in.), and the width was measured to be 48.86 mm (1.963 in.) both with the vernier caliper. The diameter of the hole was measured to be 19.9 mm (0.786 in.), and the distance from the edge of the hole to the end of the bar was measured to be 10.0 mm (0.395 in.) both with the vernier caliper. This pendulum is illustrated as Figure 1 in Attachment 1.

Measurements of the Period

The period was measured by timing 50 cycles with the wrist chronometer. The results of four such tests are given in Table 1. This investigation measured the start and end of the timing interval when the leading edge of the pendulum passed a small LED light source.

Table 1. Data for Four Each Fifty-Cycle Measurements

test	time	period
	sec	sec
1	48.16	0.9632
2	48.13	0.9626
3	48.25	0.9650
4	47.76	0.9552
	average =	0.9614
	std dev =	0.0043
uncertainty of data =		0.014
uncertainty of average =		0.007

Processing and Analysis

With the dimensional data presented above, the theoretical period of the pendulum was computed to be 0.9575 sec using the specific formula presented by Anderson (1959). Based on the uncertainties of the dimensional data as presented above, the uncertainty of the theoretical period is estimated to be no more than .0014 sec. by the method of Kline and McClintock (1953).

As seen in Table 1, the average period, the standard deviation, and the uncertainty of the period data have been computed. The uncertainty of the data, U_{data}, due to random variation has been computed using the relatively large coverage factor of 3.2 that applies in this case. The uncertainty of the average has been computed using the following standard formula from error propagation analysis,

$$U_{avg} = \frac{U_m}{\sqrt{N}} \tag{1}$$

As seen in Table 1, the result of the 50-cycle tests is an average period and corresponding uncertainty of 0.961 ± .007 sec.

Discussion and Conclusions

The results of the theoretical estimate and the three experimental measurements are shown in Figure 1. The experimental period is only 0.4% higher than the theoretical period, and the theoretical period is well within the experimental error margin. It is concluded that the experimental measurement is in good practical and statistical agreement with the theoretical estimate.

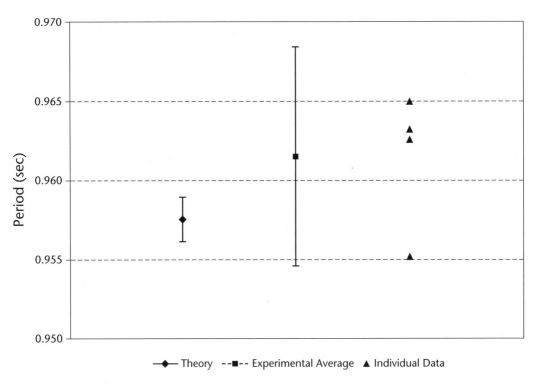

Figure 1. The Theoretical Period and the Period Measured Experimentally

Despite the good general agreement, it is apparent that the experimental measurements tend to be higher than the theoretical estimate. In fact, except for one unaccountably low datum the experimental average would be significantly higher statistically. The theoretical estimate could be too low for at least two reasons. There could be errors in the input dimensional data, or the estimated period could be low because the theory is deficient or incomplete. The dimensional measurements seem to be very simple and reliable. In addition, there is little chance of appreciable bias error in the experimental measurements of the period because the timer is very accurate; therefore, any difference probably has a physical basis not accounted for in the simple model. Since the theoretical model ignores friction, one obvious possible physical cause of the difference is the small but non-zero friction at the pivot. Additional experimental data could be taken to reduce the random uncertainty in the measurements and resolve this question.

References

Anderson, J. L., 1959, "Approximations in Physics and the Simple Pendulum," American Journal of Physics, vol. 27, pp. 188–189.

Kline, S. A. and F. A. McClintock, 1953, "Describing Uncertainties in Single-Sample Experiments," Mechanical Engineering, vol. 75, pp. 3–8.

Timex, 1996, "Instructions: Timex Quartz Digital Watch," Timex Corporation, Middlebury, CT, available online at <http://www.timex.com/instructions/086Instr.pdf>

Attachment 1

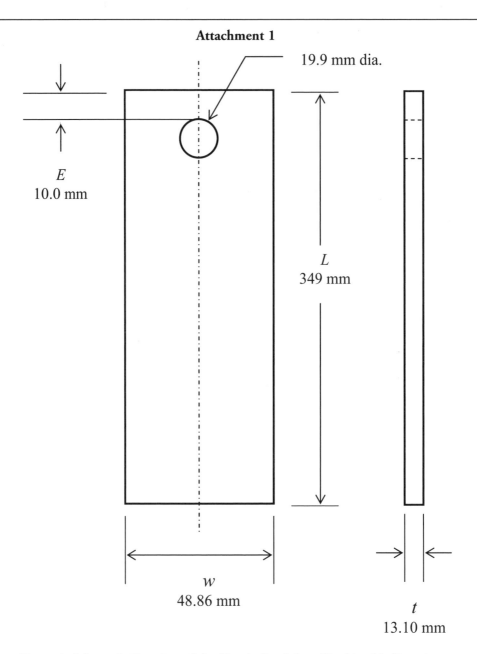

Figure 1. Schematic Drawing of the Simple Pendulum Used in this Experiment
Note: not to scale

CHAPTER 1.7

EXAMPLE OF A LONG NARRATIVE REPORT IN LETTER FORMAT

Investigation of an Oscillating Water Jet in Air
by L. C. Elwell and S. M. Jeter
for ME 4053 Engineering Systems Laboratory, 10 January 2000

Introduction. The stability of oscillating planar liquid jets in air or vacuum is important for critical applications such as first wall protection in proposed inertial confinement fusion reactors. To assess this stability under one typical operating condition, the large scale kinematics of an oscillating planar water jet were observed in the Thermal Hydraulics Laboratory of the Georgia Tech School of Mechanical Engineering on 8 November 1999. In particular, the maximum transverse deflection of the jet was measured at various positions below the nozzle. These deflections were found to agree well with predictions from a simple ballistic trajectory model.

The observations were performed using the oscillating jet facility, shown schematically in Figure 1. The jet is generated by water flowing from a rectangular nozzle, which is 1 cm wide by 10 cm long. The nozzle is oscillated by a Scotch yoke mechanism, driven by a variable speed DC motor. A typical view of an oscillating jet is shown in Figure 2. The magnitude of the oscillation can be adjusted by changing the eccentric on the motor shaft, and the frequency can be adjusted by altering the DC voltage. In this particular experiment, the zero to peak oscillation amplitude was 5.4 mm, and the frequency was 6 Hz.

The transverse deflection was measured by imaging the jet at various positions below the nozzle with a consumer VHS camera operating at the standard 30 frames per second (FPS). The nozzle exit velocity was determined using a single axis LDV (Dantec Type 55X Modular LDA). This exit velocity agreed well with the average velocity calculated from the volumetric flow.

The width, W, of the jet at the various positions downstream was first determined by imaging the jet with the nozzle stationary. The video image was then displayed on the monitor. The width of the jet image was measured directly on the screen. The ratio of the screen size to the actual size had already been determined by a direct calibration, allowing the necessary correction. The maximum transverse deflection of the jet was then determined by imaging the oscillating jet. This video was then displayed frame by frame, and the points of maximum deflection of both edges of the jet were identified. The distances between these points were measured directly on the monitor screen and corrected using the calibration ratio to give an accurate value for the width, E, of the envelope of the trajectory of the jet. The maximum deflection, δ, of the jet was then calculated by the simple formula,

$$\delta = \frac{E - W}{2} \tag{1}$$

This process was repeated for each downstream position monitored. The deflections calculated from this data should most closely correspond to motion of the centerline of the jet.

With the jet oscillating at 6 Hz and a zero to peak amplitude of 5.4 mm, the maximum deflections were determined at five positions ranging from the nozzle exit to 0.9 m downstream. The resulting data are shown in Table 1 along with corresponding predictions from a model based on a simple ballistic trajectory for the jet. Data supporting this table are given in Attachment 1.

Table 1. Experimental Data and Corresponding Model Predictions

y	E	W	exp δ	error	model δ
m	mm	mm	mm	mm	mm
0.0	21	13	4	0.7	5
0.2	34	14	10	0.8	12
0.4	51	15	18	0.9	20
0.6	65	16	24	0.9	28
0.9	90	17	36	1.6	39

As is evident in the table and even more apparent in graphical presentation, Figure 3, the experimental data agree well with a simple ballistic model. In every case, the experimental data are below the model prediction; however, the upper limits of the error limits approach the predictions. This marginal disagreement could be due to some higher order dynamic effect, or it could be due to the difficulty in detecting the very transient maximum deviation of the jet when imaged at only 30 FPS. Overall, the results show that the oscillation of a planar liquid jet in air is quite stable and that such jets can be useful in critical applications.

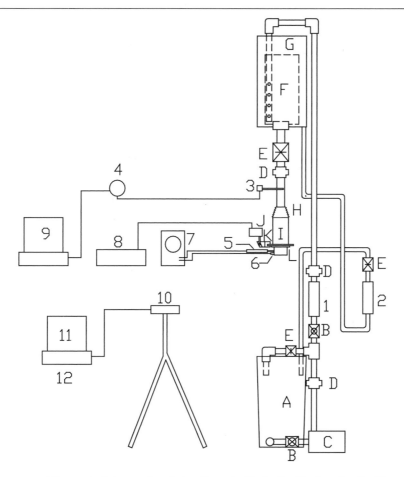

Figure 1. Schematic of the Georgia Tech Oscillating Jet Facility.

	Functional Components		Instrumentation
A	Reservoir Tank	1	Flowmeter, 0-80 GPM
B	Ball Type Flow Control Valve	2	Flowmeter, 0-5 GPM
C	Centrifugal Pump	3	Pitot Tube
D	Pipe Union	4	Pressure Transducer
E	Butterfly Type Flow Control Valve	5	Linear Variable Displacement Transducer (LVDT)
F	Constant Head Tank	6	Accelerometer
G	Overflow Tank	7	Storage Oscilloscope
H	Rubber Bellows	8	Variable DC Power Supply
I	Flow Conditioner Assembly	9	Data Acquisition System
J	DC Oscillator Motor	10	Camera
K	Scotch Yoke Mechanism	11	Video Monitor
L	Nozzle	12	VHS Player/Recorder

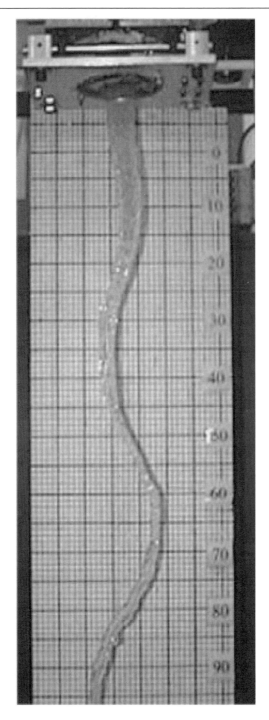

Figure 2. View of Oscillating Jet in Typical Operation
Note that the VHS camera used produced only a low resolution image.

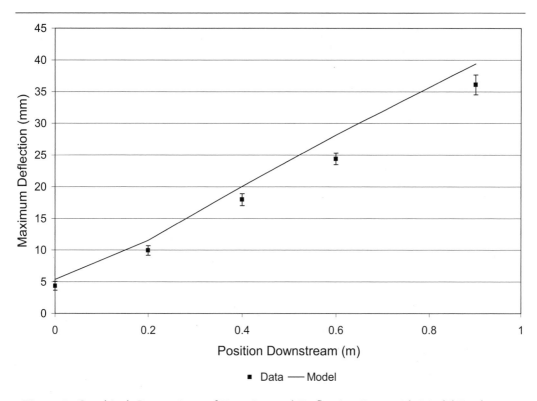

Figure 3. Graphical Comparison of Experimental Deflection Data with Model Predictions

Attachment 1. Experimental Deflection Data and Model Predictions for Planar Liquid Jet

file: OscJet00   SMJ, 6 Jan 00
Experimental data and simple model for envelope of oscillating jet.

Parameters and unique data

59 = measured flow rate (GPM)
3.80 = LDV measured nozzle exit vel (m/sec)
3.72 = calculated nozzle exit vel from FM(m/sec)
3.80 = nozzle exit velocity used (m/sec)
6.00 = frequency (Hz)
37.70 = angular velocity (rad/sec)

0.0037 = meas flow rate (m^3/sec)
0.0010 = nozzle cross-section area (m^2)
9.81 = grav accel (m^2/sec)

0.00537 = amplitude (m)

Deflection calculated from ballistic trajectory

y	t	th-max	x-0	vel-0	model delta
m	sec	rad	m	m/sec	m
0.00	0.0000	1.5708	0.0054	0.0000	0.0054
0.20	0.0495	0.4922	0.0025	0.1784	0.0114
0.40	0.0939	0.2754	0.0015	0.1948	0.0198
0.60	0.1345	0.1947	0.0010	0.1986	0.0278
0.90	0.1902	0.1386	0.0007	0.2005	0.0389

Deflection calculated from experimental data

y	E	E	W	W	exp delta	STD E	STD E	STD W	STD W	2 x STD delta
m	in.	m	in.	m	m	in.	m	in.	m	m
0.00	0.837	0.021	0.493	0.013	0.0044	0.012	0.00030	0.025	0.00064	0.00070
0.20	1.328	0.034	0.545	0.014	0.0099	0.020	0.00051	0.022	0.00056	0.00076
0.40	1.993	0.051	0.577	0.015	0.0180	0.022	0.00056	0.030	0.00076	0.00094
0.60	2.548	0.065	0.627	0.016	0.0244	0.030	0.00076	0.019	0.00048	0.00090
0.90	3.530	0.090	0.688	0.017	0.0361	0.054	0.00137	0.031	0.00079	0.00158

Comparison of Deflection Model with Data

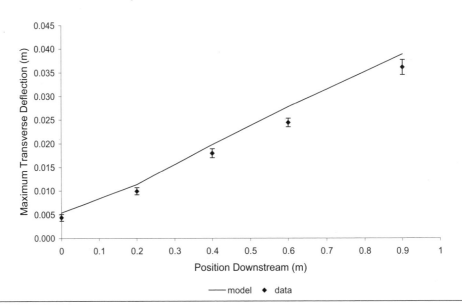

CHAPTER 1.8

EXAMPLE OF A LONG NARRATIVE REPORT IN MEMORANDUM FORMAT

TO: Instructor   23 November 2002

FROM: C. C. Pascual and S. M. Jeter

SUBJECT: Measurement of Heat Leak from the Copper Cylinder Used in Convection Heat Transfer Experiment

Introduction

The objective of this project was the evaluation of the conduction heat leakage from the heated cylinder used in the convection heat transfer experiment located in the Thermal Systems Instructional Lab of the George W. Woodruff School of Mechanical Engineering. The copper cylinder is heated by an embedded electric resistance and cooled by a cross flow of air. The power delivered to the resistance heater can be accurately measured; however, some heat will escape from the cylinder by radiation or by conduction either through the supports or through the power and instrumentation leads. This extraneous heat transfer is called a heat leak, and this heat leak should be subtracted from the electric power to yield the convection heat transfer rate.

On 9 and 10 November 2002, the heat leak from the cylinder was measured at three different air speeds by low power steady-state calorimetry. The heat leak was found to be relatively large and to depend on the air speed. A regression model for the heat leak conductance was developed that can be used to correct the measured electrical power for conduction losses. The experimental heat leak was found to be higher than expected from one-dimensional conduction through the supports. The higher value can be attributed to conduction through the electrical leads as well as the supports. In addition, the supports and one lead are exposed to the air stream and apparently act as fins accounting for the dependence on wind speed. The overall result is a regression model for a heat leak coefficient that can be used to compute an improved estimate of the true convection heat rate.

Procedure and Apparatus

Preparations

The basic procedure was to insulate the cylinder against convection, to apply a relatively small electric power to the heater, and to measure the steady state temperature of the cylinder. At steady state, it can be assumed that all the input power is escaping through the conduction heat leak paths. The heated cylinder was first installed in the induced draft wind tunnel (Aerolab Subsonic Wind Tunnel). The wind tunnel has a constant speed fan and a bypass gap to adjust the air speed through the test section. To eliminate convection and to block the radiation heat leak, the heat transfer surface of the cylinder was then covered with two layers of 25 mm (nominally 1 inch) thick pre-formed glass fiber pipe insulation. With the convection surface well insulated, almost all of the steady state power can then be assumed to be conducted away by the structural supports of the cylinder and by the power and instrumentation leads. The power leads were then connected to an electromagnetic wattmeter (Weston AC and DC Wattmeter model 310) and on to the autotransformer used to adjust the AC voltage applied to the resistance heater. The average surface temperature of the cylinder is measured by fifteen type T thermocouples embedded near the surface. These thermocouples were connected to an external submultiplexer board (Metro Byte EXP-16), which in turn was connected to an internal data acquisition board (Metra Byte DAS-8). The data acquisition board includes a high resolution 12 bit analog to digital converter. One additional type T thermocouple, also connected to the data acquisition system, was used to monitor the free stream air temperature. Finally, a portable thermal anemometer (TSI Velocicalc model 8350, serial number 291) was used to measure the free stream air speed.

Data Collection

Once preparations were complete, the wind tunnel fan was started, and the air speed was adjusted using the bypass gap to one of three representative speeds (*i.e.*, approximately 14, 9, and 4.5 m/ sec). Then, the resistance heater was energized at a relatively low power, either 5 or 10 Watts. The cylinder surface temperature was then monitored until a steady state was achieved. The experiment was hindered by the long time, 2 to 3 hours, required for the insulated cylinder to reach a steady state. Once a steady state was achieved, then the average surface temperature, the free stream temperature, and the free stream air speed were measured and recorded.

Data and Findings

The collected data and subsequent calculations are provided on the accompanying spreadsheet, Attachment 1, and the pertinent data and results are summarized in Table 1 and illustrated in Figure 1. As indicated, Table 1 includes the heat leak conductance computed as shown in Equation (1),

$$UA_L = \frac{\dot{W}_E}{T_S - T_\infty} \tag{1}$$

where,

UA_L = heat leak conductance (W/C)
$\dot{W}_E$ = electrical power (W)
T_S = surface temperature (C)
T_∞ = free stream air temperature (C)

Table 1. Heat Leak Data and Calculated Conductance

Air Speed	Electrical Power	Air Temp	Surface Temp	Experimental UA_L	Regression UA	Error
m/sec	W	C	C	W/C	W/C	%
13.9	5.0	19.0	34.3	0.327	0.331	1.4
9.0	5.0	16.6	36.2	0.255	0.258	1.1
9.0	10.0	17.4	54.5	0.270	0.257	−4.6
4.5	5.0	18.1	45.0	0.186	0.191	2.7

The tabulated experimental conductances show two notable features. First, the conductance increases with air speed. This behavior is not compatible with simple one-dimensional conductance. A 76% increase in the conductance is associated with the overall 210% increase in air speed. This increase is more typical of convection from a fin rather than mere one-dimensional conduction to an ambient heat sink. Second, the heat leak is rather large, on the order of 20% of the convection heat rate from the cylinder. The heat leak is large in an absolute sense probably as the result of excessively conductive supports and, especially, the leads. The relatively small convection heat rate, caused by the short length and correspondingly small convection surface of the test cylinder, exacerbates the relative significance of the conduction heat leak.

The experimental bias was estimated by error propagation analysis following the general method of Kline and McClintock (1953). The details are given in the spreadsheet included as Attachment 1. The result is an estimated limit of bias of ± .021 W/C. This value is less than 9% of the mid range value of the heat leak conductance.

A linear regression model for the heat leak conductance data as a function of wind speed was prepared and is shown along with the experimental data in Figure 1. The model is expressed in dimensional terms by Equation 2 for the wind speed, V, in m/s,

$$UA_L = 0.1244 \, \text{W/C} + \left(.01483 \frac{\text{W/C}}{\text{m/s}} \right) V \tag{2}$$

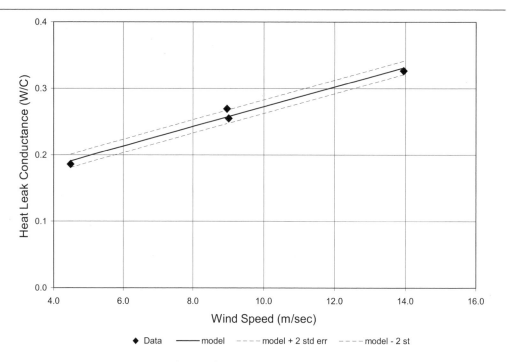

Figure 1. Heat Leak Conductance Data, Model, and Error Band

As is evident in the figure, the model represents the data well. The largest deviation of the model from the data is only –4.6%. Overall, the R-squared is almost 98% indicating near perfect agreement between the data and the model. The model has an alpha risk of only 1%, implying a negligibly small probability that the observed correlation between the conductance and the air speed could be due to mere chance. This alpha risk is well below the conventional upper limit of 5% and indicates that the regression model is statistically significant. A quadratic model was also attempted. As detailed in the attachment, this model returned only a slightly higher R-squared but has an entirely unacceptable alpha risk of 54%. This high alpha implies that the additional fit is most likely only an adjustment to random error in the model and indicates that the quadratic term is statistically insignificant and should not be used. The linear regression analysis also returned a value of only 0.010 W/C for its standard error of estimate. To illustrate the tightness of the regression estimate as implied by this standard error, an experimental error band with the half-width of two standard errors of estimate is also plotted on the figure. Obviously, all of the experimental data are either within or very near this error band, another indication that the model represents the data well and that the data includes no suspicious outliers.

Comparison with Simple Models and Discussion

The experimental heat leak conductance can be most easily compared with a simple one-dimensional conductance through the supports. The supports are made of acrylic polymer. One is a cylindrical solid, and the other is a hollow cylinder to accommodate the electrical leads. Table 2 gives their dimensions and their computed conductances. Together the two supports have a one-dimensional conductance of no more than .0073 W/K, which is less than 4% of the smallest observed heat leak.

Table 2. Simple Conductance Calculations for the Solid and the Hollow Supports

	Units	Solid	Hollow	Sum
Inner Dia	mm	0	32	
Outer Dia	mm	32	44	
Area	mm^2	792	760	
Length	mm	41	0	
Conductance	W/C	0.0038	0.0034	0.0073

To explain the much larger observed heat leak conductance, it was noted that the cylinder is equipped with 16 pairs of 24 gage thermocouple leads, one pair of 12 gage power leads, and a 12 gage grounding conductor. A single 12 gage copper lead effectively 50 mm long has a conductance of .026 W/C, and a 24 gage lead has a conductance of .0016 W/C. In total, the electrical leads could allow a heat leak conductance of at least .104 W/C, which is 54% of the lowest observed value. In addition, the supports and the grounding lead are directly exposed to the air stream and are sure to experience some lateral conduction and ultimately convection heat loss. Furthermore, the power leads are connected directly to the resistance heater, which is much warmer than the convection surface itself, exacerbating the heat leak by this path. These effects could easily account for the enhanced conduction heat leak, and the fin effect especially explains the unexpected dependence on the wind speed.

Radiation Heat Leak

In addition to the observed convection heat leak, in normal operation, the cylinder is cooled by radiation. Since the cylinder is not provided with a low emittance coating, its radiation heat loss is not negligible. For nominal temperatures of 50°C surface and 25°C ambient and assuming 0.3 surface emissivity, the radiation heat leak is calculated to be around 0.04 W/C. The calculated radiation heat leak is somewhat more than 15% of the nominal conduction heat leak. This additional leak is large enough to be significant and should also be subtracted from the input power when computing the convection heat rate.

Closure

The conductance heat leak from the forced convection test cylinder was measured and a regression model was developed. For the expected range of wind speeds the heat leak conductance is significant, ranging from 0.19 to 0.33 W/C. In addition, the radiation heat leak can be estimated to have a nominal value of about 0.04 W/C, also an appreciable contribution to the heat transfer from the cylinder. Proper correction of the input power for the conduction and radiation heat leaks should significantly improve the accuracy of the forced convection experiment.

Reference

Kline, S. A. and F. A. McClintock. 1953. Describing Uncertainties in Single-Sample Experiments, *Mechanical Engineering*, 75: 3-8.

Attachment 1. Spreadsheet for Heat Leak Experiment

file: FCCAL3   20 Nov 98 (SMJ)
Experimental determination of heat leak from forced convection source
Data collected 8 November 1998.

Summary:

0.010 = alpha risk	0.021 = error limit (W/C)
0.010 = R-squared	8.123 = percent error limit

Constants and parameters:
0.3048 = m/ft
1609.3 = m/mile

Data and calculations:

speed FPM	speed m/sec	power W	t-amb C	t-sur C	UA W/C	model UA W/C	error %	model + std err	model - std err
2745	13.9	5.0	19.0	34.3	0.327	0.331	1.4	0.341	0.321
1773	9.0	5.0	16.6	36.2	0.255	0.258	1.1	0.268	0.248
882	4.5	5.0	18.1	45.0	0.186	0.191	2.7	0.201	0.181
1763	9.0	10.0	17.4	54.5	0.270	0.257	-4.6	0.267	0.247

Linear Regression output:

Constant	0.124391
Std Err of Y Est	0.010105
R Squared	0.979721
No. of Observations	4
Degrees of Freedom	2

X Coefficient(s)	0.014833
Std Err of Coef.	0.001509
t-stat	9.829773
alpha	0.010191

Quadratic Regression output:

Constant	0.092897
Std Err of Y Est	0.010752
R Squared	0.988519
No. of Observations	4
Degrees of Freedom	1

X Coefficient(s)	0.022650	-0.000422
Std Err of Coef.	0.009073	0.000482
t-stat		-0.875413
alpha		0.542230

Uncertainty analysis:

nom Ts =	36.2	C
nom Ta =	16.6	C
nom W =	5.0	W
nom UA =	0.255	W/C

	u	(units)	dUA/dx	(u dUA/dx)	(u dUA/dx)^2
T (air)	0.733	C	0.0130	0.0095	9.11E-05
T (surface)	0.189	C	0.0130	0.0025	6.07E-06
Power	0.063	W	0.0510	0.0032	1.02E-05

u in UA	0.0104	W/C
error limit	0.0207	W/C
error limit	8.1	percent of nominal UA

CHAPTER 1.9

GUIDE TO LONG REPORTS

Introduction

Experimental projects in capstone courses are larger in scope and more challenging than the projects in other undergraduate experimentation courses. In the sample report included in this chapter, for example, the students were asked to modify a fluid flow system in such a way that it would produce stable laminar flow and turbulent flow, as desired. They were then asked to evaluate the modified system experimentally to determine whether it really did produce stable and turbulent flow, as desired. To meet these goals, the students evaluated the existing system in order to design appropriate modifications. They then implemented a modification, and they tested the modified system to verify that the modification had been successful.

Because capstone projects are complicated, you will have more information to report than has been the case in previous laboratory courses. Consequently, you will occasionally need to modify the report format you have used in your other lab classes. The main format headings of experimental project reports are essentially those of the experimental project letter report; however, as the sample report shows, you will need to squeeze in some extra information from time to time. We have developed this guide to help you to insert such additional information responsibly, clearly, and concisely. An assembly checklist precedes the sample report.

The format details that follow are designed to be suggestive and flexible. We suggest that you follow this format as you begin developing your reports, and that you then modify this format as required by the specifics of your projects. The following discussion illustrates one way to address the issues of a complicated project using the established headings of a lab report.

The Body of the Report
Introduction
This section should describe the functional objective of the project, it should supply any background information pertinent to the problem, and it should raise the specific technical issues or questions that you need to address during the project.

Apparatus
This section should indicate what apparatus was used, and what modifications were introduced to any existing systems; it should also describe what, if anything, was designed or fabricated for use during the project.

Procedures

As in any experimental report, this section should describe the different tests that were conducted.

Data

As in any experimental report, this section should briefly present the experimental data, using exhibits as needed.

Analysis

As in any experimental report, this section should briefly describe how the results were evaluated, drawing conclusions as needed for each analysis operation. Uncertainty analysis should be presented here.

Closure

As in any experimental report, this section should briefly review the project's tasks and concisely describe the important findings. New information should not be introduced in this section of a report.

Production and Editing

Page and paragraph formatting, exhibits, and production qualities are important elements of professional presentation. In addition to the general guidelines presented above, consider the following guidelines on format, exhibits, attachments, and editing when you prepare your reports.

Regular paragraph style

Student reports should use the semiblock paragraph style, which has two characteristics: a single blank line is inserted to separate paragraphs, and the first line of each paragraph is indented. In addition to these characteristics, you should observe the general standards for titles, headings, and subheadings. Indented subheadings for paragraphs, called *run-in side heads*, should be underlined or bolded, they should be followed by a period, and they should be immediately followed by the text. Headings should appear alone on a line, they should be bolded, and they should be aligned on the left margin. A heading should follow a blank line, and text should begin on the next line. Main titles should be bolded, centered, and set off from surrounding text by blank lines above and below.

Spreadsheets

If a spreadsheet is required for the report, include as an attachment a printout of all important parts of the preliminary spreadsheet. The spreadsheet should be as complete and tested as possible. The spreadsheet should have a unique number and name (or caption); this information should be centered and prominent on the page. Use the header utility to generate a title similar to that shown here:

Attachment 1. Experimental Spreadsheet

Exhibits

The specifics of your work will be described in figures, tables, spreadsheets, pictures, and the like. When you attach such material to a report, or when you integrate it into the text, then it becomes an exhibit. Exhibits must have numbers and descriptive captions, and they must be inserted or attached in the order that you will direct the reader's attention to them. Spreadsheets use a prominent, centered header as a descriptive caption. Table numbers and captions are placed above the tables, while figures are numbered and captioned below the figure.

Figures and tables should be incorporated into the text of your reports. When an exhibit such as a spreadsheet is not appropriate to be integrated into your report, you may attach it at the back of the report.

Appendices

Use appendices or other attachments for long mathematical derivations or calculations. As a minimum, appendices are generally expected to include a page of sample data, a page of sample data analysis, and a page of sample uncertainty analysis. Attach appendices after the text.

Attachments

Any exhibits or appendices attached to the report are still part of the report. Their format and production qualities should match those of the report. Every attachment must be cited in the text, and every attachment must have a unique number and a descriptive caption.

Final editing

Carefully proofread your report. Do not rely exclusively on your spell checker.

Assembling the report

A long report should be assembled with a cover page showing the title of the project, the names of the students on the team, the course and section numbers, and the date of the report. A long report should also have a table of contents, which is followed by an abstract and then the body of the report. At the back of the report, an acknowledgements page should follow the conclusions. References, attachments, and appendices are then added respectively.

Assembly checklist for Experimental Project Reports
A Quick Reference Sheet

An experimental project report includes, but is not limited to, the following main elements:

Cover sheet

Table of Contents

Abstract

Introduction
> *Objective*
>> What were you asked to do, in functional terms?
>
> *Background*
>> What is known about the system you will use?
>
> *Problem*
>> What technical questions must you solve?
>> (Only needed to the extent that the functional goal does not specifically cover this matter.)

Apparatus
> What equipment/materials/instrumentation were used?
> What modifications were introduced to existing apparatus?
> What, if anything, was designed (controllers, pumps, test stands, and so forth)?

Procedures
> What tests were run to validate the design, innovation, or modification?

Data
> What data were collected?

Analysis
> What operations, if any, were performed on the raw data?
> What does the data reveal about the system?

Closure
> Does the system perform as expected?
> What issues remain open?

Acknowledgements

References

Appendices

The Generic School of Mechanical Engineering
Standard Institute of Technology, Collegeville GA 30332-0405
Undergraduate Instructional Laboratories

TO: Undergraduate Laboratory Instructor   11 April 2002
FROM: Undergraduate Laboratory Students in Group A03
SUBJECT: Letter of Transmittal of Final Report

Dear Professor Burdell:

Transmitted herewith is the final report of our session long project on the development of an improved water flow system for the LDV experiment in the undergraduate lab. We thank you for your encouragement and assistance throughout this project.

Sincerely,

Ernst T. Clark, Group Leader

Wileng A. Scholar

Claire L. Scribner

Attachment: Final Report

ME 4055

FINAL REPORT

CONSTRUCTION AND VERIFICATION
OF A SYSTEM FOR RELIABLE
LASER DOPPLER VELOCIMETER MEASUREMENTS
IN LAMINAR AND TURBULENT FLOWS

GROUP A03

Ernst T. Clark, Group Leader

Wileng A. Scholar

Claire L. Scribner

ADVISOR: J. P. Burdell

10 April 2002

The Generic School of Mechanical Engineering
Standard Institute of Technology
Collegeville, Georgia 30332-0405

TABLE OF CONTENTS

Abstract

An improved and very stable fluid flow system is needed to properly demonstrate laminar flow in an undergraduate fluid dynamics laboratory. This project was undertaken to satisfy this need. The particular objectives of this project were:

1. To modify an existing fluid flow system to allow it to reach stable turbulent flow and stable laminar flow through a circular glass tube.
2. To use a Laser Doppler Velocimeter (LDV) to characterize the fluid flow through the modified system.

The existing system was modified through isolation of the pump from the existing system by introducing a constant head tank. In addition a variable area flow meter with a modulating valve was installed. The modification was so successful that the modified system produced steady flows at Reynolds Numbers as low as 300 to 400. The modified system is also capable of demonstrating stable turbulent flows at Reynolds Numbers up to 23000. Most importantly, the modified system is able to produce stable laminar flows for Reynolds Numbers as high as 1300. For Reynolds Numbers above 1300, the system exhibited transitional behavior of the type usually expected at Reynolds Numbers around 2100.

Introduction

The objective of this project was to design, construct, and test a very stable water flow system. This flow system was required to produce stable laminar and turbulent flows in a circular glass pipe over a wide range of volumetric flow. This flow system was to be developed as a modification of an existing fluid flow system that could not be regulated well enough to develop stable laminar flow. The existing system was modified to eliminate flow variations caused by close coupling the water pump. A Laser Doppler Velocimeter (LDV) specifically a TSI System 9100 then successfully measured fluid velocity profiles over a wide range of flow rates.

Background

A schematic of the previously existing LDV and flow system is shown in Figure 1 below.

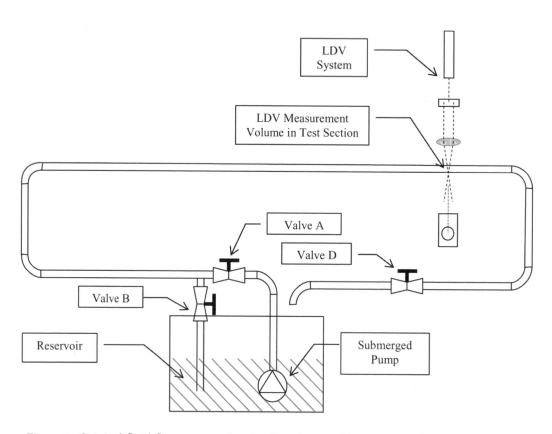

Figure 1. Original fluid flow system, showing locations and integration of LDV System, test section, reservoir, pump, and flow control valves

The flow variations of this system stem from the close integration of the pump, which was connected directly to the pipe loop containing the glass measurement pipe. This direct connection caused erratic velocity profiles, particularly at the low Reynolds Numbers typical of laminar flow. The primary variations appear to be flow oscillations resulting from dynamic interaction between the pump and piping system. The pump further disrupted the fluid flow by introducing mechanical vibrations into the measurement pipe and the return tube. The existing flow control valve also presented some problems. The valve may have contributed to turbulence in the fluid flow, and it was difficult to precisely adjust low flow rates using the existing valve.

To address these problems, the pump was isolated from the system in a series of modifications shown in Figure 2. A constant head tank and an overflow tank were added to the system between the pump and the measurement pipe. This modification eliminated pump surge by allowing for a consistent fluid head, which is provided by the difference of fluid levels in the head tank and the reservoir. In design of this system, the height of the head tank required careful consideration. This tank had to overcome head loss due to frictional forces from the pipelines, valves, and fittings throughout the system. However, the height of the head tank was constrained by the limited pump head. Additionally, the return tube was a possible source of fluctuation, and the system had no flow meter to allow direct reading of the flow rate.

Fluid velocity in this system is measured in the glass measurement pipe downstream from the pump reservoir. Measurements are taken with an LDV, which crosses laser beams through the glass pipe at a point where the flow is fully developed. The LDV, shown in Figure 3, on the following page, operates on the differential Doppler principle. A monochromatic helium-neon laser beam is sent through a beam splitter to produce two coherent beams. These beams are then directed by a lens to cross inside the circular glass measurement pipe. When these beams cross, they produce a pattern of light and dark fringes that are separated by a known distance D_f. As entrained particles cross through each bright fringe, a pulse of scattered light is focused into a mirror, which reflects the light pulse into a photomultiplier tube (PMT). A data processing unit connected to the PMT detects the fringe passing frequency from these pulses. The velocity of the fluid is calculated from the known distance, D_f, between the fringes and the measured frequency.

A flowmeter was needed to allow direct measurement of the flow rate of the liquid. In response, a rotameter or variable area flowmeter was installed. This meter verified the calculated flow rate, thus validating the collected LDV data. The rotameter itself can be calibrated by the weighing tank method as described below. Future experiments can compare the volumetric flow calculated from LDV data with direct measurement of the flow using the rotameter.

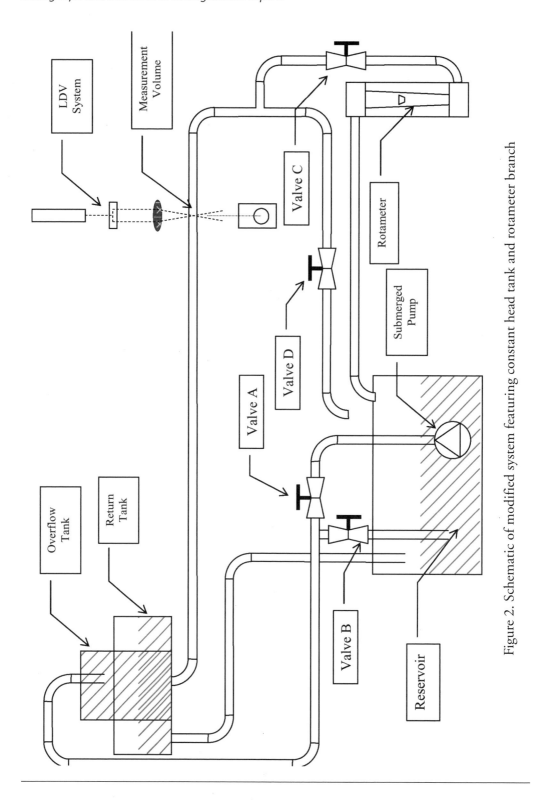

Figure 2. Schematic of modified system featuring constant head tank and rotameter branch

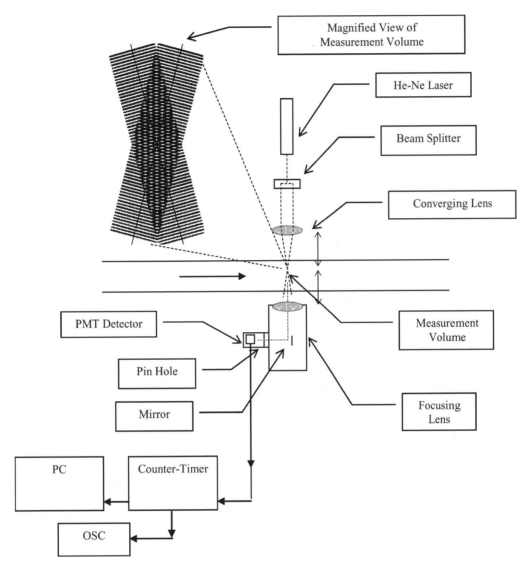

Figure 3. Schematic of the Laser Doppler Velocimeter, showing the He-Ne laser,
beam splitter, photomultiplier tube (PMT) detector, and signal processing components

Apparatus and Procedure

The modified fluid flow system is shown in Figure 2 above, which shows the new position of the pump, head tank, and overflow tank, as well as the return valve, or adjustment valve, and the flow meter. The system is started when the pump is activated, sending water from the reservoir tank into the constant head tank. The water flow into the head tank is regulated through adjustments to the bypass valve and the throttling near the reservoir. Overflow from the head tank is captured in the overflow tank, from which it returns to the pump reservoir. Water flows from the head tank into the glass measurement tube, where its velocity is measured with the LDV. The water then continues through the return pipe to the reservoir tank.

Flow profile. To determine the maximum velocity of the water flow, the LDV must measure the velocity profile across the glass measurement pipe. This profile is started at a near wall reference point calculated to be about .43 mm (0.017 inch) from the tube wall. To locate this point, the laser traversing table is moved toward the near tube wall, taking measurements until a low particle count is obtained. A dial indicator attached to the adjusting table is then set to zero, and the adjusting table is moved a distance of 27.9 mm (1.09 inch) from the first point, which should place it in the desired position about 0.43 mm from the far wall. The data rate is noted, and the procedure is repeated until similar rates are counted at both sides of the tube, indicating that a symmetric alignment has been found. Once this adjustment is completed, the LDV measurement volume should be centered so the flow rate can be adjusted.

The LDV output to the computer, as shown in Figure 3, is used to measure the maximum velocity of the water at the center of the tube. The flow rate is adjusted to the desired velocity using the flowmeter throttling valve. The pump throttling and the bypass valves are adjusted to a flow that just overflows the overflow tank without reducing the flow through the pump to an unacceptably low rate.

The laser is then reset to the near wall location .43 mm from the tube wall, and the average velocity, standard deviation, and turbulent intensity are measured. This data is entered into a spreadsheet to be used for further analysis. Once the data for a point has been recorded, the laser is moved 1.3 mm (0.050 inch) toward the other wall, and more data is taken. This data collection is repeated until measurements have been taken across the whole of the tube. At that time the volumetric flow and the actual Reynolds Number are calculated in the spreadsheet, and the velocity profile is plotted.

Calibration of flow meter. The flow meter on the right in Figure 2 was introduced to the system to independently determine the volumetric flow rate of the fluid in the measurement pipeline. This flow meter must be calibrated so the flow meter scale will correlate to known flow rates. This calibration was performed using the weighing tank method. A digital stopwatch and a truck scale, specifically a Toledo Type 211795 scale, were used.

For calibration, the flow meter was set up to flow water into the weighing tank. The scale of the flow meter is read as numbers from 0 to 100 in increments of 10. When fluid runs through the flow meter, a float inside indicates a number on this scale to indicate the fluid flow rate. The weight of the water inside the tub changes as fluid leaves the flow meter and enters the tub. This changing water weight is recorded at the beginning and end of a time interval lasting between 3 and 5 minutes. This process is repeated for each desired increment on the scale of the flow meter.

The resulting change in mass in kilograms of water is converted to a change in volume in liters and then divided by the corresponding change in time in seconds to give a volumetric flow rate. The measured values of volumetric flow rate (L/sec) were plotted against the scale numbers of the flow meter. Regression analysis gives this calibration equation,

$$Q_{corr} = 0.0418X + 0.044 \frac{\text{liter}}{\text{min}} \tag{1}$$

where Q_{corr} is the volumetric flow rate in L/min and X is the scale number. The R^2 value for this calibration is 0.9994, and the Combined Uncertainty of the calibration is 0.21 L/min. These results are completely acceptable for this application. This equation can be used to determine the corrected volumetric flow rate that corresponds to the scale number indicated on the flowmeter.

Results

The objective of this project was to design a fluid flow system capable of producing stable laminar flow as well as stable turbulent flow, and this objective was accomplished. With the addition of the head tank, the system now produces a laminar parabolic flow profile for the first time, as shown in Figure 4. The flow velocity profile in the figure for $Re = 390$ closely resembles the laminar model based on Equation (2),

$$u/U_c = 1 - \left[\frac{r}{R} \right]^2 \tag{2}$$

$$U_c = 2u_{ave}$$

Some error is still apparent. Poor adjustment in the measurement locations may have caused some shift in the measurements. However, the data in Figure 4 still give adequate but not completely identical agreement with the model. Even with the residual error, the modified system is a great improvement over the previously existing system, which never produced even an approximately parabolic flow profile.

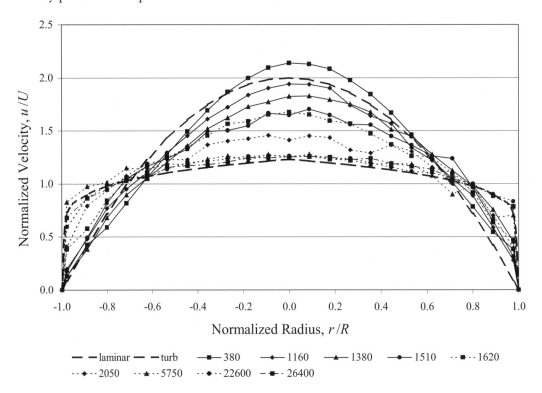

Figure 4: Flow profiles obtained with the modified system over a range of Reynolds Numbers

The range of this laminar flow for the modified system was also determined. As the speed of the fluid approaches the transition zone (*i.e.,* Reynolds Number 2100 to 4000), the flow profiles grow less stable, yielding curves that are not smooth, as shown in Figure 4. For this system, the velocity profile begins to show some fluctuation or instability as low as Re = 1500. This early apparent transition to turbulence is probably due to some disturbance induced in the flow by the piping system and some residual turbulence in the fluid when it reaches the test section. Consequently the effective range for this system is below 1500. The obvious challenge is to upgrade the system so that it can generate a smooth laminar profile for Reynolds

Numbers closer to 2100, the textbook measure of the end of the laminar zone; however, this enhancement would certainly be non-trivial.

The nature of the transition from laminar to turbulent flow was found to be more interesting and complicated than usually appreciated. The most characteristic features of this transition are the normalized centerline velocity and the corresponding standard deviation of the velocity. Obviously, the normalized centerline velocity decreases during transition as the profile evolves into the more uniform turbulent profile. As tabulated in Table 1 and illustrated in Figure 4, the normalized velocity in the core flow does tend to decrease as the Reynolds Number increases. This behavior is as expected; however, the decrease is substantial even within the accepted laminar range as presented in all standard texts such as Munson, Young, and Okiishi (1998, p. 477). During the tests, the display of the active histogram of the local velocities fluctuated when the flow rate was relatively low. The results were the uneven flow profiles obtained in Figure 4. Fluctuation is expected in the transition range, but some minor fluctuation was observed in real time and demonstrated in the profiles that were collected for Reynolds Numbers beginning at 1200, well below the expected 2100.

The monotonic decline in the normalized velocity as the Reynolds Number increases is a characteristic feature and deserves some further investigation. According to theory, the normalized centerline velocity should remain at 2.0 in the laminar range, which corresponds to a Reynolds Number somewhat below about 2100. In this regime the velocity profile should be parabolic. After transition to turbulence around a Reynolds Number of 4000, it should remain nearly constant at around 1.2 corresponding to the expected power law profile. In the transition region (*i.e.,* $2100 < Re < 4000$), there should be a decline in peak velocities. Figure 5 shows measurements of the peak velocities plotted against the logarithm of the ascending Reynolds Numbers. In the transition region, the normalized peak velocity decreased as expected.

The trend in the standard deviation is not as widely documented as the trend in the normalized velocity, and its behavior was found to be more complex. The normalized standard deviations of the centerline velocities are shown by the error bars on each point in Figure 5. This statistic is direct evidence of scatter in the velocity data caused by turbulent fluctuations. The scatter tends to increase as the Reynolds Number increases in the nominally laminar regime. However, rather than continuing to increase, the normalized standard deviation reaches a maximum in the transition region where the normalized standard deviation is larger than in either the laminar or the highly turbulent regions. This behavior, while not necessarily expected, is reasonable. In the transition region, the onset of turbulent eddies causes relatively large velocity fluctuations. This fluctuation increases into the mildly turbulent range. However, as the flow rate is increased even further, the velocity in the turbulent eddies clearly does not increase as quickly as does the local mean velocity. Consequently the standard deviation decreases and the ratio called the turbulence intensity even begins to decrease. The greatest relative fluctuation is then in the transition not the fully turbulent regime.

Table 1. Normalized velocity and normalized standard deviation at the centerline of the test section for the experimental range of Reynolds Numbers (*Re*)

Re	U_c/U_{avg}	norm SD at CL[a]	*Re*	U_c/U_{avg}	norm SD at CL[a]
384	2.14	0.0342	2050	1.41	0.0291
1160	1.94	0.0458	2710	1.37	0.130
1380	1.82	0.0398	5750	1.25	0.159
1510	1.65	0.0501	22600	1.26	0.104
1620	1.68	0.0322	26400	1.25	0.0949
1820	1.71	0.0493			

[a] The normalized Standard Deviation (SD) of the velocity sample at the pipe centerline (CL).

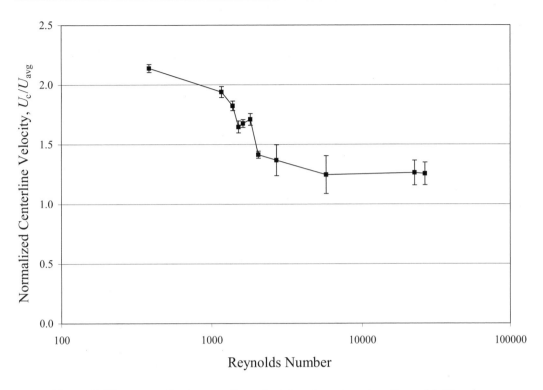

Figure 5. The normalized centerline velocities plotted against Reynolds Number
accompanied by error bars 2 sample standard deviations in length
to illustrate the relative turbulence

Discussion

Overall performance of the modified system is acceptable. However, at higher Reynolds Numbers but still within the expected laminar zone, the flow profile becomes unstable and deviates from the model. Initially, this problem was thought to stem from back pressure fluctuations in the return line to the reservoir. In the original modification, the outlet of the return line was below the water level. This arrangement allowed unavoidable fluctuations in the reservoir water level and currents in the reservoir to effect flows in the test section. To eliminate this possibility, the pipe exit was raised so that water pours freely into the reservoir at ambient pressure. Some improvement in the flow stability was noted after this improvement, but fluctuations were still not eliminated.

The design of the constant head tank may be a source of this remaining disturbance. In order to maintain a constant head without splashing, the water must be discharged below the surface of the water in the tank. This submerged inlet may currently be so close to the exit of the tank that fluctuations in the discharge may influence the flow in the test section. In support of this hypothesis, air bubbles were observed to sometimes collect in the measuring tube. These bubbles are created by splatter from the return line at the reservoir; and they then can be induced into the pump and travel through the supply line to the head tank. At low flow rates any bubbles sent to the head tank should rise to the top and leave the system. However, even at relatively low Reynolds Numbers in the upper end of laminar flow range, some bubbles remain in the liquid long enough to be carried into the measuring tube. Indeed, the tube must be flushed of bubbles periodically. This phenomenon is exacerbated at higher Reynolds Numbers because at low laminar Reynolds Numbers the flow is slow enough to allow bubbles to escape from the free surface rather than being drawn into the measuring tube. Obviously a larger constant head tank would be desirable, but available space limits installation of a much larger tank.

The fluctuations might be minimized by moving the discharge of supply line to head tank as far as possible from the drain without actually lifting it out of the water. This modification should be feasible. Alternatively, a diffuser could be attached to the hose entering the head tank. This diffuser would allow fluid to flow more smoothly from the pump into the head tank, preventing the pumping surges from directly disturbing the flow in the test section.

It is also possible that the flow profile has not fully developed by the time it is measured. The required entry length, l_E, which is the length of pipe necessary for full development of the velocity profile, is quoted in standard texts such as Munson, Young, and Okiishi (1998, p. 464) as

$$\frac{l_E}{D} = 0.06 \, Re \tag{3}$$

The measuring pipe is 3.0 meters long, and its diameter is 38.1 mm. Consequently when the flow rate exceeds that necessary for Reynolds Number to be 1300, the current pipe is too short for laminar flows to develop fully. The simplest way to solve this problem is to use a measurement pipe with a smaller radius. A longer pipe could also solve this problem; however, space constraints make such a pipe impractical.

Closure

An existing fluid flow system was modified to allow it to reach stable turbulent flow and stable laminar flow through a circular glass tube. A Laser Doppler Velocimeter was used to characterize fluid flow through the modified system.

The modified system produced Reynolds Numbers as low as 300 to 400, it produced stable turbulent flows at Reynolds Numbers up to 23000, and it produced stable laminar flow for Reynolds Numbers as high as 1300. For Reynolds Numbers above 1300 the system exhibited transitional behavior of the type usually expected at Reynolds Numbers around 2100.

Several recommendations have been developed to address the problems of premature transitional behavior at Reynolds Numbers between 1300 and 2100: (1) the size of the head tank should be increased, (2) a diffuser should be added to the supply line, and (3) the supply line discharge into the head tank should be moved farther from the outlet. In addition, the diameter of the glass tube should be reduced to insure that the water has achieved fully developed flow at the measurement point.

Acknowledgements

We would like to thank our instructor for providing us this opportunity to work directly with Laser Doppler Velocimeter technology and to apply our knowledge of fluid mechanics and fluid systems to a real life application. We would also like to thank our teaching assistants for steering our progress in the laboratory and for providing us with much needed advice and materials. They also instructed us on how to use the laboratory equipment and ensured that our work was correct.

Reference

Munson, Bruce R., Donald F. Young, and Theodore H. Okiishi. 1998. *Fundamentals of Fluid Mechanics,* 3rd Ed., John Wiley & Sons, Inc., New York.

Appendix: Sample Calculations for
Determination of Reynolds Number from Average Fluid Velocity

The Reynolds Number is related to area average fluid velocity according to the following equation:

$$Re = \frac{VD}{v}$$ (A.1)

where: Re = Reynolds Number
V = Average fluid velocity
D = Fluid passage diameter, pipe inner diameter in this case
v = Kinematic viscosity

Representative experimentally obtained values for these variables were:
V = 0.0512 m/s   (0.168 ft/s)
D = 38.1 mm   (nominal 1 1/2 inch glass pipe, 1.50 in. ID)

From a standard reference the kinematic viscosity at 25°C representative of experimental conditions is
v = 9.03 ′ 10-7 m2/s   (9.72 ′ 10-6 ft2/s)

Substituting these values into Equation A.1 gives

$$Re = \frac{0.0512\frac{m}{s}\,0.0381m}{9.03 \times 10^{-7}\frac{m^2}{s}} = 2160$$

Other Reynolds number calculations were similar.

CHAPTER 1.10

GUIDE TO ORAL-VISUAL COMMUNICATION

Introduction

Oral presentations of technical results are very like the written reports that might be used to present the same work. Oral presentations use the same information format as do written reports, they display the same graphical information, and they make the same points. However, presenters must attend to three practical problems in preparation and delivery of their information:

- *Brevity*
  Presentations are commonly limited to five or six minutes; consequently, presenters must be careful to select only the most valuable information to include in their talks.
- *Display*
  Text and graphics must be displayed on a screen whose visual requirements are different from those of a text page; presenters should expect to develop special copies of graphs, drawings, and tables in order to use screen space effectively.
- *Balance of text and graphics*
  Text information can be delivered in the form of text on a slide and in the form of spoken words; presenters are encouraged to make points verbally and to minimize the words that are placed on slides.

The following discussion will focus mainly on the issues of display and balancing text with graphics. A sample presentation follows the discussion.

The Slide Sequence of an Oral Report

Cover

The first slide of an oral presentation is typically a text slide that does the same job as does the cover sheet of a written report. It displays the name of the report, the name of the author/presenter, that person's affiliation, the date, and any other required information. Slide 1 of the sample presentation is a rough characterization of a cover slide.

Overview

The second slide of an oral presentation is typically a text slide whose job resembles that of the abstract in a written report. It briefly characterizes the project and the results that will be presented. Slide 2 of the sample presentation below exemplifies the appearance of an overview slide.

Objectives

The third slide of an oral presentation is typically a text slide used to characterize the objectives of the work that is being described. Complete sentences are to be avoided in this slide; it is frequently sufficient to list the forces or phenomena that are to be measured, analyzed, characterized, or determined. An example would be the third slide in the sample presentation.

Apparatus and Procedures

One or two slides may be used to characterize the tools and methods involved in data collection. Slides 4 and 5 of the sample presentation might both be used in a description of an experimental project.

Data

As in written reports, data in oral presentations is commonly presented in tables and graphs, and there might be many of these in a talk. Slides 6 and 7 of the sample presentation stand in for the many graphical slides that might be used to package and display experimental data.

Analysis using equations

Experimental results are generally numerical, and it is important to show the equations that you use to arrive at your numerical result. Slide 8 indicates how an equation is best displayed on a slide.

Professional appearance of slides

Slides 1–8 are devoted to the correct presentation of technical information using slides. Slides 9 and 10 offer general tips for legible and useful presentation of words and sketches on slides.

Presenting yourself professionally

Slides 11–13 provide simple suggestions that student speakers should review during the last few minutes before delivering a presentation. These suggestions involve both quality-assurance review of slides and reminders for deportment during the presentation.

Closure

As does the final section of a written report, the final slide of a technical presentation should provide a simple summary of the points that have been made over the course of the talk.

Production and Editing

Presentations are made using projectors. Some of these are quite good, and others are very poor, and it is not always possible to discover in advance whether you will be provided with a good projector or a poor one. The following suggestions for slide production are designed to make sure you have a good result even if your projector is poor.

Font size and selection

Font sizes should be kept between 20 point and 40 point. Fonts smaller than 20 point will not reliably display, and fonts larger than 40 point will waste too much space on the slide. Arial is the preferred font for most presentations, as it projects better than do serif fonts such as Times New Roman.

Text density

Words should be kept to a minimum, even on slides that describe project goals. Complete sentences should be avoided in favor of lists of key terms. For example, in an Objectives slide, one might reasonably use verb-noun statements such as these

- Measure
- Determine
- Characterize
- Develop

Professional quality graphics

Projected graphics need to have a clean and sharp appearance. While errors can sometimes be overlooked in small print graphics, these errors will be expanded to huge proportions by a projection system. It is thus best to prepare drawings, tables, graphs, and equations using computer drawing and graphing tools.

Font levels

It is usually unnecessary to give your words extra emphasis through use of visual markers such as boldface, italics, and the like. Audiences generally find such visual markers to be distracting. Presenters should avoid heavy use of such visual markers; instead, speakers should create emphasis using such fundamental presentation tools as their voices and their pointers.

Colors and backgrounds

In most cases, color does not add information to a slide. Poorly chosen colors can accidentally obscure important information, and even the best color decisions are quickly lost if a black and white photocopy must be made. To avoid this problem, you should follow these guidelines for slide design:

- Background colors should be uniform, light, and dull. White is an excellent background color for informational slides.
- Foreground colors, such as those in drawings and graphs, should be dark. Black is an excellent color for the lines in technical drawings and graphs.
- Distinctions are best shown by variations in line weights, line breaks, and the like.

In short, you should exhaust your black-and-white visual resources before you begin to integrate colors into your graphics.

Horizontal space

A projection screen provides a great deal of horizontal space, and you should strive to fill that space with information. Drawings and graphs should be sized to fill the screen from left to right; when a graphic does fill the horizontal space on the screen, several blank lines of vertical space will usually remain above and below the graphic. This space is best reserved for slide titles, placed above the graphic, and for captions and legends, to be placed below the graphic.

Fonts on graphs and tables

Most graphs and tables are prepared using one piece of software, such as Matlab or Excel, before they are placed in a second piece of software, such as Powerpoint. Unfortunately, the font sizes originally selected for the axis labels and legends may not survive the transition, leaving these labels illegible when projected during a talk. To complicate this problem, there is no easily available table of font translations which will help the student to select, say, a Matlab font size that will scale to 20-point during projection. Very simply, the student must conduct a few calibration tests with each piece of graphing software and each piece of presentation software, setting legends, labels, and titles to 20-point, and then adjusting by a few points until the graph or table projects legibly. While this process is labor-intensive, it can be dispatched quickly and relatively painlessly, and it must be repeated each time a new piece of software is integrated into the author's toolkit.

Graphs and tables

Data can typically be presented as either a graph or a table. However, graphs and tables do different jobs for the speaker and for the audience; thus, it is usually not necessary to present all data in both forms, and authors must decide which form of presentation is appropriate for a given data set or for a given project. In making such a decision, authors must consider the strengths and weaknesses of each form of data display. Tables permit great specificity, as they offer a simple and detailed view of data and of the operations that are performed on that data. However, some people find it difficult to spot trends in tabular displays of data, and most people find it difficult to track and compare trends in multiple data sets when they are presented. Graphs support comparisons between sets of data, benchmarks, and the like, although in doing so, they often sacrifice specificity, as it is generally impossible to determine the X and Y coordinates of a given point with visual examination of the plot.

In professional notebooks, graphs and tables are likely to be kept together. For quick display in presentations, it is best to avoid tables in favor of graphs. It is also wise, however, to place important tables on slides and keep those tables handy during the presentation; because these tables provide comprehensive views of data, they can provide necessary details during the question-and-answer session that follows every presentation.

Photographs

When an apparatus must be illustrated, students should provide drawings rather than photographs. When a photograph is used in a presentation or report, that photograph should be aligned with a drawing that schematically represents the important elements of the device or apparatus of interest. In such a circumstance, the photograph should be used in support of the drawing, as professional drawings are generally more powerful and valuable tools for engineers. Additionally, photographs raise two simple problems that most technical professionals cannot solve alone: it is very difficult to take a technical photograph that isolates your device from surrounding clutter, and it is very difficult to manage lighting so that your device displays fully in a photograph.

Speaking for figures

After the slides have been developed, it is still the job of the speaker to describe those slides to the audience and to explain what point is made by each graph or drawing. While work on the presentation begins with development of good slides, the presentation itself still depends on the words that the speaker says. No speaker should assume that a slide's point is obvious to the audience; you must always go on record by stating the points that your slides illustrate. Those slides do not speak for you; rather, you must speak for those slides.

Slide 1

DESCRIPTIVE
TITLE

NAME

AFFILIATION

Orientation:
Constrained to portrait?
Is landscape feasible?

type size $\geq$ height / 20

Slide 1 shows an outline for a title slide, showing the relative locations and sizes of the descriptive title, author's name, and institutional affiliation. Text lines are typically centered on title slides, and names and titles are typically presented in capital letters. Type size should be 20-point or greater.

As on the cover sheet for a written report, the descriptive title of the report is placed toward the top of the sheet, followed by the author's name and the name of his/her company, and the presentation date. Students should expect to include extra information, such as the course and section numbers and the instructor's name; the specific items to be included will be specified by the instructor.

This slide raises the question of orientation. Instructors and professional audience members do not mind whether students prepare slides in portrait or landscape orientation so long as the slides are consistent. It is a very bad practice to alter the slide orientation during a presentation; audiences may be tolerant of such variation, but speakers often find that they lose track of the orientation changes.

Most presentation programs offer landscape-oriented slides as a default, and these work well for technical documents, as technical graphics tend to be wide. However, projection systems vary, and many old-fashioned overhead projectors are designed for portrait-oriented transparencies. Students should find out what kind of projection system they will use during their presentations; when such information is not available, it is wise to leave large margins on all sides of each slide.

SLIDE 2

Outline:

1. Introduction:
Define the Investigation or Analysis
Motivate the Audience

2. Describe the Experiment or Theoretical Approach

3. Present the Data and Results or Findings

4. Summarize and Conclude

The second slide of a presentation should present an outline of the information that will be presented in the talk. This outline slide, as represented by Slide 2, essentially lists the major format headings and indicates what information will be provided in each section of the talk.

Slide 2 should not be copied directly; however, it is a reasonable generic prompt for the outline of any experimental talk. Note that each line of Slide 2 begins with a verb form, urging the student to take an action: *Define, Describe, Present, Summarize*. When speakers fashion their outline slides, these lines must show the results of these actions:

- This experiment investigated the relationship between . . .
- The investigation was conducted using . . .
- The data show that the relationship is . . .

Students should present this information in roughly 15 to 30 seconds.

Objective

Stated Concisely

(Devote an early slide to a concise
statement of the objectives of the project.)

Slide 3 represents a statement of experimental objectives. The project objectives should be stated concisely and specifically. In making this statement, students should avoid showing complete sentences on their slides. Instead, students should display a set of verb-noun phrases to characterize the different tasks of which the project is comprised. The verbs will characterize the things that experimental professionals do, such as measure, characterize, develop, design, determine, analyze, test, or verify. The nouns specifically name the technical concepts, tools, or ideas that are analyzed, verified, used, tested, determined, or developed. Students need to present this information in about 30 seconds.

SLIDE 4

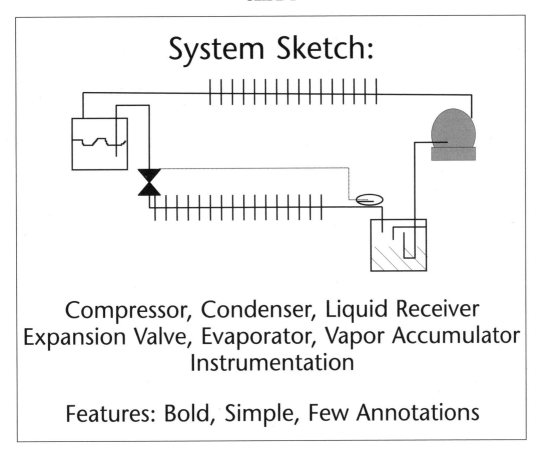

System Sketch:

Compressor, Condenser, Liquid Receiver
Expansion Valve, Evaporator, Vapor Accumulator
Instrumentation

Features: Bold, Simple, Few Annotations

Slide 4 shows a simple sketch of a refrigeration system; a sketch of this quality and detail can be used effectively to convey the general layout of an experimental apparatus. Simple sketches of this sort are helpful because they strip away unnecessary detail in order to convey large structures; if finer views are necessary, they can be presented in a subsequent detail illustration. In general, descriptions of experimental systems should isolate those structures that are of interest, such as structures that move, that are adjusted, or that respond to inputs. A slide such as this should be described in 30 to 45 seconds.

SLIDE 5

Use photo of apparatus
only if accompanied by schematic.

Slide 5 presents a photograph of the experimental apparatus that is sketched in the fourth slide. Photographs of this sort should never be presented unless they are accompanied by a simplifying sketch. Indeed, photographs are generally to be avoided in student reports and presentations because they present practical problems:

- Laboratory lighting is often poor, obscuring the sharp distinctions that are the objective of technical illustrations;
- Photographs capture unnecessary details, which are often more prominent than the structures that are of experimental interest;
- To overcome the first two problems, photographs require extensive labeling, which can obscure information as well as highlight information.

Excellent photographs can be obtained and used in support of experimental reports if the student has the money and time to consult with a professional photographer. Those students who lack such resources should rely on technical sketches to convey information.

Concise Tables

Station	Temp °C	Press kPa	Enthalpy kJ/kg
1	8.9	393	192
2	50.9	965	211
3	39.5	948	74
4	8.5	405	74

Avoid complex tables or raw spreadsheets.
Display Significant Digits Only!
Include critical or sample data only.

Slide 6 represents the appearance of an appropriate table as it might be placed on a presentation slide. Of interest here are the table's size, simplicity, format, and its management of numbers.

Size

This table has only twenty cells, including column heads. Because there are few cells, the type size can be kept at 20-point or greater, the minimum for visibility on a presentation slide.

Simplicity

Simplicity is obtained through careful selection of the information that is to be presented and through arrangement of that information in support of a point. The table shown in Slide 6 is not comprehensive and does not strive to be so. Rather, tables such as this are created by carefully selecting crucial bits of information from a much larger table of experimental data. In presentations, good tables commonly summarize the information that is captured in much larger and more comprehensive experimental spreadsheets.

Format

The cells in this table enclose the numbers and words tightly; very little space is left above and below the numbers. The numbers themselves are separated by thin, dark border lines around the cells. Each column has a complete, brief header label, showing units when they are available.

Numbers

The columns of numbers are carefully aligned here. The decimal points are vertically aligned in the *Temp* column, and the values are right-aligned in the *Enthalpy* column. Further, the display of significant digits is both consistent and credible.

A slide such as this should be described in the space of about 30 or 45 seconds.

SLIDE 7

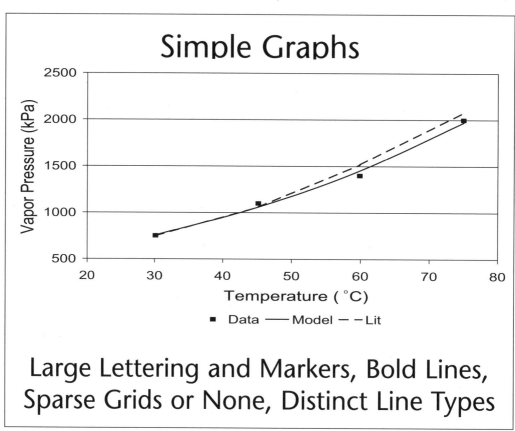

Slide 7 represents the appearance of an appropriate graph as it might be used in a presentation. Of interest here are the graph's size, its use of lines and markers, the appearance of the background and grids, and the display of axis labels and legend.

Size

The graph in Slide 7 is sized to fill the slide from side to side. This large display makes it easier for audiences to read the labels and legend, which are provided by the graphing software rather than the program used to build the presentation. The plot is further sized to spread the data across the whole of the display; only small empty areas appear in the regions to the left of 30 degrees on the X-axis and above 2000kPa on the Y-axis.

Lines and markers

Each mark on a graph is meaningful, so all the marks on the graph must be visibly distinguished by manipulation of line weights, breaks, and marker shapes. Information on this graph is represented in three distinct forms. Small squares, called markers, represent the experimentally obtained data. The smooth line represents a model, calculated by the students using experimental data and pertinent equations. The dashed line represents a model that is defined in the literature; the student-obtained data and the student-developed model are compared to this literature model. A legend at the bottom of the graph defines what the different lines and markers represent.

Background and grids

Data is best observed when it is displayed with dark lines and markers arranged on a light background. The white background of this graph provides excellent visibility for the black lines and markers on this graph. This graph also demonstrates that grid lines are largely unnecessary; three Y-axis grids extend lightly across the plot area; X-axis values are indicated only by small cross-lines that extend below the plot area. Such grids are appropriate because graphs are not designed to be used as look-up charts, a job best left to tables. Rather, graphs are best used as tools to display and relate sets of data and predictions.

Axes and legends

In this graph, as in all good graphs, both axes are fully labeled, and units are shown in each label. A legend is also clearly visible, providing a key to the different markers and line weights that are used on the plot.

A slide such as this should be described in roughly 30 or 45 seconds.

Legible Equations

$$COP_H = \frac{\dot{Q}_{COND}}{\dot{W}_{COMP}}$$

No tiny type even in scripts
Use the size menu in Equation Editor

Slide 8 represents the reasonable and appropriate display of an equation. Of interest here are two very simple points. First, equations need not be massed together on slides. Generally, only a few equations are important to a project, and only these important equations need to be shown in a presentation. Second, the equation here is large; it is sized such that the smallest subscripts print out at roughly 20-point size, which again is the smallest character size that can reliably and effectively be used in a presentation slide. Students should describe a slide such as this in roughly 30 seconds or less.

SLIDE 9

Overall Slide Design

Avoid clutter

Few concepts per slide

Key words and phrases only

No long sentences or text

Slide 9 provides simple, summary tips on effective slide design. Clutter in technical slides is best avoided by designing figures and charts according to the principles described above. Text slides can be kept free from clutter by sticking to key words and phrases only. Complete sentences are to be avoided on all slides; after all, the speaker needs to talk during a presentation. A speaker needs to be prepared to add value to the visible information on the slides, and he or she needs to make simple and direct points about each of the slides.

Slide Quality

No hand drawn figures or hand lettering

No tiny type

(even for notations, equations)

No multiple fonts

Slide 10 makes the simple point that slides should look professional; hand-prepared slides demonstrate to the audience that you are not yet a professional. All the type that appears on a slide should be larger than 20 point in size. Further, all speakers ought to avoid mixing fonts, typefaces, and emphasis on slides. The slides, important as they are, are only the visible part of the presentation. The speaker is best positioned to create emphasis using pointer and voice.

SLIDE 11

Preparation

Rehearse and Time, Respect Time Limit
Allow for questions, interruptions

Check Spelling and Drawings

Check for Internal Numerical Accuracy

Sanity Check Overall Results

Preview Presentation Room

Verify Equipment / Supplies

Slide 11 speaks to issues of preparation, and these require that speakers get help from their colleagues and/or team members. Presentations should be rehearsed aloud, with one student tracking time while another student speaks. Students should have their friends review their slides and illustrations in search of obvious errors, such as the misspelling on this sheet. Students should have their team members review their slides, in search of numerical inaccuracies and/or flaws in preparing equations, table heads, graph labels, and so forth. And, finally, speaker(s) should strive to examine the room where a presentation will be delivered, to determine whether the projection equipment works, how it works, and whether the slides that have been prepared will really work on the equipment in that room.

Speak Effectively

Avoid colloquial terms.

Use professional, technical language.

Descriptions should be precise
and quantitative.

Speak clearly the last sound in each word
and the last word in every sentence.

DOn't drOp yOUr cOnsOnants.

Slide 12 provides some fundamental tips for presenters. Very simply, presentations should be taken seriously, and speakers are expected to deliver talks that are formal, specific, and complete. Speakers should adopt the specific, direct, and concrete language that best describes professional technical experimentation. Speakers should pronounce their words fully and clearly, and speakers should finish their sentences at full volume.

SLIDE 13

Presentation Style

Dress appropriately.

Guide the audience.

Do not block projector.

Use pointer not fingers.

Avoid excessive informality.

Express a definite ending.

Slide 13 offers specific suggestions for the speaker's conduct during a presentation. Speakers are obliged to respect their audiences. This means that they should be aware that audiences are sensitive to whether the speaker's dress is appropriate and that audiences require guidance from the speaker. Speakers further should be sensitive to the audience's visual problems; speakers should avoid obstructing the projector, no matter where it is placed, and they should use pointers. Last but not least, audiences require a clear indication that the talk has finished; speakers can provide this mark by saying "Thank you for your time. I'll be happy to answer questions."

Part Two:

Standards for Undergraduate Reports

Sheldon M. Jeter

CHAPTER 2.1

INTRODUCTION

Producing an engineering report is a design and production project just like any other engineering project. The business of composing a report may not be as interesting as the research or development that it documents; however, the report can be equally important because innovation requires change, and change will not happen unless inventions and discoveries are communicated. This second part of this text on engineering reports discusses the job of producing a report. Here, some ideas will be presented about applying the tools used in engineering projects—planning, design, and implementation—to the preparation of a technical report.

Plenty of successful projects in every field are completed without any apparent planning or organization. The participants just rely on luck and hard work. A lot of time and effort will probably be wasted, but if the project is successful, the waste will probably be forgotten. The same can be true of the minor task of producing the report that documents the project, but here the wasted time and effort and the possibly less successful outcome seems particularly unfortunate for at least a couple of reasons. First, the engineer should be a professional with respect to design, research, and development, but he or she is probably an amateur when it comes to report preparation. The seasoned professional can rely on experience and ingenuity, but the amateur should employ planning and standard methods. Second, the engineer should save his or her energy and resources for the most valuable work he or she can contribute, which is engineering, not writing. So, consider an engineering report to be an engineering project that can be done much more efficiently and effectively by using the engineering method.

Every engineering project, like almost everything—except maybe a Mobius strip—can be seen to have a beginning, a middle, and an end. Most projects are organized at least implicitly around the three corresponding stages of conceptual design, preliminary design, and detailed design. In the conceptual design, the goal is defined and the scope and general features of the design solution are envisioned. In the preliminary design, the major components are identified and integrated into a cohesive system. In the final detailed stage of design, the construction documents and detailed specs are generated. The same organization works well for report writing. This part of your text is organized around these three stages.

The first stage of conceptual design is definition and invention. The analogous stage in report writing is defining the type and scope of the report that is required or suitable, the appropriate overall design or format, and the general content or outline. The next chapter addresses these issues.

The next stage, or preliminary design, is when the project engineer exults and excels. This is the stage of selecting the components and putting them into functional subsystems, then connecting the subsystems into effective systems, and ultimately integrating the subsystems into an optimal whole. The analogous stage in report writing is developing the first draft. This is the time to flesh out the outline with well structured paragraphs of effective sentences that present the technical content. Most of the technical content will actually be presented in exhibits such as graphs and tables. The following section in this part addresses the general design of these components and subsystems.

The final stage of design is putting the finishing details in order and assuring good quality control in writing and production. The final sections of Part II of this text are devoted to the details of these important issues.

CHAPTER 2.2

BASIC STANDARDS AND COMPOSITION

Introduction

This section is concerned with conceptual design issues. These issues include defining the type and scope of report that is required or suitable, the appropriate overall design or format, the general content or outline, and the basic standards for technical writing.

Technical reports fit into generally distinctive standard types, and these types are typically presented in a few standard formats. Similarly, good technical writing follows some standard conventions on compositions. A familiarity with report types will help the writer identify the correct type for a particular kind of technical communication. Familiarity with formats gives the writer a preview of the ultimate product. Finally, familiarity with standard technical composition gives the writer the knowledge to use the available tools and objects to reach the desired goal of efficiently producing an effective report. In this section, the basic standards on report types and formats are first considered. Once the general overall goal is described, some general guidance on the composition needed to implement the reports will be given.

Report Types and Formats

Technical reports include abstracts, itemized reports, and narrative reports. Each type has been developed to fit a different role. All the types will be encountered in the typical undergraduate experimental engineering curriculum and in engineering practice. The types, their roles, and their corresponding general structures are reviewed briefly below.

Abstracts

Abstracts are essentially stand alone reports about longer reports. Short abstracts can be a single paragraph or a few paragraphs that are used to accompany and define a more complete report. This kind of abstract can also be entered into a database to catalog the subject report. A longer abstract will resemble a short report and is often called an executive abstract. The title describes its intention. An executive has general responsibility, so an executive abstract should inform a person with general responsibility and knowledge, but not specific involvement, about the results or status of a project. A special blend of short report and abstract is the so-called extended abstract that is used to accompany and document some oral and visual presentations.

Itemized Reports

Sometimes test results and exercises are reported using a rather structured itemized format. Such documents report only some specified items about a test, exercise, or experiment. Itemized reports resemble routine industrial test reports.

Narrative Reports

Longer reports are usually narrative reports. The term narrative report is used in this text to identify and describe a report that tells the entire story or narrative of an experiment or project.

Length

If no word or page limit is specified, the author should use only as much space as is necessary. Technical abstracts should be one—or at most a few—paragraphs. Sometimes, a definite word or even character limit is imposed on abstracts. Executive abstracts can be a couple to several pages long. Typically, shorter narrative reports should not exceed a few printed pages, exclusive of figures and tables. The author must strictly comply with any page limits established by an editor or instructor. In student reports especially, do not pad the report. A lack of understanding will rarely be successfully concealed by a jumble of words. Do not dwell on insignificant issues or echo routine details of standard procedures. When discussing the apparatus and procedures, define the items concisely without excessive reiteration of insignificant details. Overall, writers should strive for a concise and pertinent report.

Report Formats

Several conventional formats are in use. A convenient form for a relatively short narrative or itemized report is a business letter or "technical memorandum." Alternatively, a short report can start with a title block of specified design. Major reports require a long format with chapters and special sections. Long format reports for major projects should have a cover page, a table of contents, some special sections, and the body of the report. Typical special sections include lists of nomenclature, figures, and other exhibits. The body should be organized into chapters or sections according to contents and functions. A common example of a long format report is one written for a semester length experimental engineering project.

The formats described here should suffice for all typical applications. In undergraduate courses, the syllabus or course instructions should specify the required report type. In professional practice, the engineer should select the appropriate type to fit the project. Once the type and format have been selected or specified, the writer should outline the report, identify the needed exhibits, and compose the report. Composition should meet the basic standards of technical writing, so these standards are presented in the following section. Following that, the objects and tools that make effective writing possible are presented.

Basic Standards of Technical Writing

Developing good technical writing style is the aim of the first part of this text. While important, style is somewhat subjective, and it certainly requires practice and editing. Not everyone will quickly develop a engaging style, but every student or practicing engineer can comply with the basic standards of composition that apply to technical writing. If the basic standards are met, then the foundation for effective style will be in place. The author can improve the writing style by further editing or by getting the help of a professional editor. With practice, less editing or help will be necessary, but the basic elements must be in place for editing to be effective.

In general, lab reports and technical literature should be primarily informational. If a technical report is effective and efficient, it will be appreciated even if the text is not inspiring or entertaining. Good technical writing needs only to be reasoned, organized, informative, objective, concise, and efficient. The organization is usually emphasized by headings that identify the topical subdivisions of the report. The chapters and sections in this document are examples. Efficient presentation requires effective exhibits. Exhibits are characteristic features of technical reports, and they include figures, tables, equations, lists, and technical attachments. Because of their critical importance, exhibits must be especially well-designed and presented. Preparing effective exhibits is the topic of several of the following chapters. General guidance can be found in Chapter 2.5, and guidance on the specific types is located in Chapters 2.8 through 2.13.

The first criterion listed above was "reasoned" because this criterion is essential. *Reasoned* means presented in accordance with logic and experiment. In experimental engineering, the immediate source of empirical knowledge is the experiment itself, but it is not the only source. Well-established basic scientific knowledge is a storehouse of logic and experiment that can be applied generally to the analysis and interpretation of experimental results. For example, the principle of conservation of energy applies universally. The investigator can apply such scientific principles generally, ensuring that only scientific principles are applied correctly. Engineering information is another resource, but it is slightly different in scope. Reliable engineering information is based on scientific principles and methods, but it is characteristically limited in range of application. When engineering information is used in a technical argument, the investigator must ensure not only that the information is formally correct, but also that the information suits the application. For example, it is blatantly incorrect to compare an engineering model developed only for laminar flow of a specific class of liquids with the findings of an experiment on the turbulent flow of a gas.

The rules and practice of logic cannot be reviewed in the available space, but one must be sure that evidence supports the argument and that the argument supports the conclusion. To be valid, an empirical argument must meet two criteria. First, it must be consistent with all relevant experimental evidence and all applicable general principles. Otherwise, the conclusion or interpretation could be easily falsified. Second, the evidence must be sufficient to exclude all other possible alternative interpretations. Otherwise, the conclusion would be incomplete, and in an engineering application, it could be too incomplete to be useful. Again, evidence can either be established knowledge or reliable experimental observation. Be aware that even correct reasoning, when based on incomplete or faulty information, can lead to an absurd conjecture. Similarly, superficially plausible, but actually incorrect, reasoning can lead to an incorrect, illogical result—even if the underlying evidence was reliable.

The conscientious author must also be on guard against some related intellectual mistakes. These might be called rhetorical mistakes because they are bad ways of developing or presenting arguments. Three such mistakes are especially common.

One shortcoming is obscurantism, which is presenting the argument or conclusion in unnecessarily complex or unusual terms. The argument may be valid, but if the wording is too obscure to be understood by the intended audience, it will be useless in engineering.

A second shortcoming is equivocation, understood to mean failing to adopt an obvious conclusion. When the argument supports one definite conclusion, it is unprofessional to throw up a screen of marginally valid but unlikely alternative conclusions.

The third error is especially common in undergraduate work: resorting to a tautology (*i.e.*, making a circular argument) when an empirical explanation is required and available. For example, a normalized pressure can be defined as the ratio of a measured dynamic pressure to some reference pressure. This value may be low in a particular region of the flow field. It is a mere tautology to state that the normalized dynamic pressure is low because the measured pressure is relatively low. This assertion, while true, is a mere tautology, a restatement of the definition with no additional insight. A better interpretation would be to observe that the normalized dynamic pressure is low in the boundary layer. While the tautology was true but devoid of insight, the alternative interpretation was true *and* added some physical insight.

Remember that creating an acceptable, or even excellent, technical report is a straightforward task and that adequate technical writing is a readily learned skill. If the report is designed with the basic standards of technical writing in mind, then proper application of the components and tools should generate an acceptable or even excellent product.

Employing the Tools and Objects of Composition

The most important objects in technical writing are sentences, paragraphs, and technical exhibits. The words and data already exist, and the overall outline is likely to be either obvious or specified, so the value added in technical writing is in drafting effective sentences and paragraphs. The tools for developing these objects are composition, editing, and proofreading.

Outline

The overall outline of a technical report will probably be fairly obvious, but the writer—and especially the student writer—is well-advised to sketch out and outline before beginning. The outline ensures that the report is suitable for the prescribed purpose, and it lays out a string of markers and associated tasks. Do not bother to make the outline a finished product on the first pass. Allow it to be a working document, continually updated as the composition proceeds. Once the tasks are completed, the report will essentially be complete. Experimental reports in particular tend to follow a specific outline, so a generic outline is presented at the end of this chapter to be used as an example or a template.

Sentence composition

In sentence composition, one should use short and complete sentences. Eliminate sentence fragments and rambling run-on sentences. Run-on sentences are easy to fix. Break run-on sentences into several simple sentences. Alternatively, try several simple sentences and one compound or complex sentence. Technical writing, like the work it describes, should be objective. To enhance the overall impression of objectivity, one should compose an impersonal narrative avoiding the first person. The passive voice helps avoid inappropriate first person references when describing a procedure or an observation. For example, the statement "we measured the vapor pressure" is unacceptably informal and personal. Instead, write "the vapor pressure was measured with a calibrated Bourdon tube gage." Otherwise, use the active voice judiciously in impersonal narratives. For example, the active construction "The temperature increases the vapor pressure" is simpler and more direct than "The vapor pressure is increased by an increase in temperature." Choosing the proper tense is not difficult. Use the past tense to describe a completed action such as the experimental procedure. The past tense is also appropriate for describing a finding of limited and perhaps even temporary validity, such as "The flow in pipeline A was found to be turbulent." In contrast, express a universal finding or fact with the present tense, such as "The Reynolds number influences transition to turbulence." Simple, logical, and objective sentences are the goal. For sentence composition, the prime directive is to avoid sentence fragments and rambling run-on sentences.

Paragraph composition

Paragraphs are the most important subdivisions for organizing and developing your ideas. In your composition, emphasize well-structured paragraphs. A structured paragraph has an over-all goal—the topic—and an organized and connected train of thoughts—the sentences—that meet this goal. An occasional short paragraph inserted for special emphasis may be acceptable. Typically, however, a paragraph should develop one topic and have three parts: 1. a beginning or introduction that presents the topic; 2. a middle, or substantive text, where the topic is developed; and 3. an ending, which is the closure or transition. The supreme practical rule in paragraph composition is analogous to the rule for sentences: avoid paragraph "fragments" and rambling "run-on" paragraphs.

Exhibits

Graphs, tables, figures, and other exhibits are critically important in technical writing. Indeed, most effective technical writing is organized around presenting the exhibits. Detailed guidelines on each kind of exhibit are provided in the following chapters, but a few general rules always apply. Every exhibit, such as a figure or a table, must be cited in the text, and authors must include every exhibit that is cited. Exhibits should have descriptive titles that include a unique, consecutive number such as "Figure 1" or "Table 1." All titles should be centered and emphasized. An important technical rule is to avoid insignificant digits in text and tables, so the rules for identifying significant digits are reviewed in Chapter 2.16. The detailed guidelines for each type of exhibit are given in a designated section below.

Assembly

Follow any mandatory requirements. Straightforward assembly and unobtrusive binding with simple covers, or even no covers, present a sober appearance. These features may be required and are always recommended. For coursework, the report should be neatly stapled in the upper left-hand corner. Use a heavy duty stapler if the report is rather thick. Report covers should not be necessary and may not be allowed. Alternative binding may not be allowed. Be sure the report is securely stapled; you are responsible for lost pages. Also, be sure the sharp ends of the staples do not protrude; why risk injuring and irritating the readers?

Editing

Final editing is analogous to the final design when the engineer ensures that the components fit and work together to meet the design requirements. A reasonable procedure is to first check the exhibits, which tend to be the most critical components. Ensure that the exhibits are correct and effective and that they are presented in the correct format. Next, check the paragraphs. Avoid paragraph "fragments" or rambling "run-on" paragraphs. Apply the tools presented in the style section to integrate the paragraphs into cohesive structures. Finally, check for sentence composition, and proofread the document.

Proofreading

Proofreading is correcting specific grammar, spelling, and other detailed errors. Grammar and spelling are critical. Use the spell checker as a zero-order check, but also personally proofread the report carefully. It is helpful to have a team partner proofread the final work. In final proofreading, the very best practice is to read the report backwards two times. The first time, read backwards sentence by sentence to check for construction and logic. The second time, read word by word to double check the spelling and usage. Common, more serious, errors are: using "to" for "too" and "it's" for "its," using an incorrect article, omitting a verb, and using subjects and verbs or pronouns and nouns that do not agree. Remember that excessive grammatical errors will detract from even a technically excellent report.

Summary Table

Table 2.1 summarizes the guidelines in this chapter, and it can be used for editing and grading general aspects of undergraduate reports.

Table 2.1. Guide to General Aspects of Experimental Engineering Reports

General	
Report is prepared in correct format, *e.g.,* title block or memorandum.	Cover page, abstract, and table of contents included if required.
Report is dated and signed.	All required or necessary items and topics are addressed.
The writing style is effective, concise, scientific, and efficient.	The report is complete and well organized.
Padded text and redundant comments avoided with no excessive length.	Jargon or colloquial writing avoided.
Conjectures and vague or subjective arguments are avoided.	All conclusions are justified on scientific and empirical basis.
Text is grammatical and proofread, not merely spell-checked.	The report is securely and safely bound.
Introduction defines project and answers WWWWH and W?*	Closure is complete, concise, and effective.

*Who, what, when, where, how, and why?

Table 2.1. Guide to General Aspects of Experimental Engineering Reports *(cont'd)*

References	
Citing and listing of references is conventional and scholarly.	Page and text formats are conventional, correct, and consistent.
Exhibits and Attachments	
All exhibits and attachments cited are included.	All exhibits and attachments included are cited.
All exhibits and attachments are of professional quality.	Named exhibits (*e.g.,* "Table 1" not "the table") are capitalized.
Units and Digits Displayed	
SI units are primary (USCS may accompany).	No insignificant digits in text or tables.
Statistics	
Adequate attention is given to any statistics and regression analysis.	R-squared and alpha risk addressed for every regression model.
Page, Section, and Paragraph Formats	
Margins are uniform on top and bottom and on both sides, usually one inch.	No wide bottom margins appear, especially at exhibits.
Section headings are used and accurately identify the role of the text.	Text in a section addresses the topic completely without misplaced remarks.
Modified block paragraph format used.	Paragraphs throughout text have clear objectives and address requirements.
Paragraph Style	
Paragraph topic sentence is present.	Paragraph topic is developed.
Extra line is skipped between paragraphs.	
Sentence Style	
A half line is skipped between every line of text.	Conventional sentence grammar is used.
Incomplete fragments or run-on sentences are avoided.	Simple sentences predominate with only isolated compound or complex sentences.
The text is impersonal.	First person is avoided or minimized.
Active voice is used only judiciously.	Parallel structure is used in list or series (*e.g.,* imperative sentences in listed procedure).
Common typo errors (*e.g.,* "too" for "to," "it's" for "its") are avoided.	Common technical writing faults are avoided.

SAMPLE GENERIC OUTLINE FOR AN EXPERIMENTAL ENGINEERING REPORT

Heading

Use letterhead or an address block.

Introduction

If appropriate, write the introduction in the style of a stand-alone abstract.

Identify the place and date of the experiment. Describe briefly the general procedure and goals. Identify but do not detail all the important or interesting findings.

Avoid presenting **fine details** of the apparatus or procedure or results in the introduction; use general descriptions here instead. Reserve details for body of report.

Do not directly cite the literature or even exhibits in the report here. For example, do not cite "Table 1" directly. Instead refer indirectly to "the tabulated results."

Background

Omit this section in shorter narrative reports.

Apparatus

Describe the general apparatus. If appropriate, include a very simple schematic sketch.

Identify the important materials, equipment, and instrumentation. Use either the "(generic) commercial" or the "generic then commercial" style of identification in the **first** instance only. An example is ". . . an electret capacitance microphone, specifically a Realistic Model 33–20, . . ." Afterwards, identify the equipment generically, for example, "the electret microphone."

In experimental reports, be sure to assess the uncertainty of all the important instrumentation that is used to make quantitative direct measurements. Identify the basis of the assessment, such as resolution, observation, or calibration; and explain how the assessment was made. Include a report of the calibration status of any calibrated instrumentation, and cite the calibration reports in the standard author (date) format.

Procedure

Describe the general procedure. If a consensus standard (e.g., ASTM or ASHRAE) technique or a technique published in the literature is used, then cite the source using the standard author (date) style.

Describe the specific steps of procedure. As appropriate use either paragraph style or list style. If used, a list must have some introduction and closing. The items in the list must be rhetorically parallel.

Describe any special precautions followed during the experiment such as monitoring conditions closely or pausing for system to equilibrate. Also describe and justify any deviations from or enhancements to the standard procedure.

Present Data

Describe and explain any preliminary routine data processing such as correction for calibration. Describe and **justify** any other adjustments or corrections.

Present the data in tabular and/or graphical form. Pay special attention to the format and quality of all exhibits. Present concise table of only the pertinent data. **Never** rely exclusively on an informal spreadsheet attachment. Report only significant digits in text and tables.

Describe the **general** trends and any important **specific** features in the data.

Compare the experimental data with literature, theoretical, or other expected data or models. Cite any literature sources for comparative information.

Present Results of Analysis

Describe the data processing or calculation method. Present the results in concise tabular and/or graphical form.

Compare results with literature or theory, citing the sources. If possible, explain the agreement or disagreement with comparative results.

Present any general results or overall conclusions based on this analysis.

Present Other Data as required.

Present Other Results of Analysis as required.

Closure

Briefly summarize the project, and briefly recap all the important findings. As desired or appropriate, use paragraph or list style. Recall that a list must have some introduction and closure, and the items should be indented. The items in the list must be rhetorically parallel, meaning that they should be presented in similar grammatical structures, such as in all noun phrases.

Do not introduce any new information in the closure.

References

Present complete bibliographical information in standard format.

CHAPTER 2.3

GUIDELINES FOR EDITING SPECIFIC FEATURES OF REPORTS

Several features are common to almost every technical report. The engineer or student who is beginning to practice effective technical writing should become familiar with and use the following editing guidelines that are applicable to these features. These guidelines represent a reasonable consensus about the use and construction of the more common and important specific features in present day technical writing.

Page Design

The most important aspects of page design are margins, justification, and line spacing. Careful attention to these features is essential to an overall visual appearance that inspires confidence in the technical content.

Margins

Uniform margins are absolutely essential to an integrated and professional appearance. Uneven margins and wide bottom margins especially give a report the appearance of being merely an assembly of unrelated notes, not a unified report. Margins should typically be one inch wide all around. Take particular care to avoid wide bottom margins. A wide margin signals the end of a chapter in a long report or the very end of any report. Anywhere else, a wide bottom margin represents a false—and apparently premature—ending that is sure to confuse the reader and damage the author's credibility. Competent technical writing should result in a report that is well-coordinated and integrated from paragraph to paragraph and from section to section, so one should ensure that the printed text as well as the ideas flow smoothly from page to page.

Line spacing

Research indicates that single-spaced text is somewhat slower to read (Kolers et al., 1981). This disadvantage probably comes from the difficulty in quickly resetting the eyes correctly to the start of the next line when the lines are too closely spaced. Professionally printed material is usually single spaced, probably to save paper, but it is set in narrow columns to facilitate resetting the eyes. Ordinary reports will be printed a full letter page wide. Here, it is advisable to always double or 1.5 space the lines. This extra space is particularly needed in student reports and in drafts so that the reviewer or grader has room to make notes.

Justification

The alignment of text is called justification. Subsection headings are usually flush with the left margin. Numbers in columns are naturally flush with the right margin. Full justified text is adjusted so that it is even with both margins. Full justification *does* disrupt the word spacings, so it might be expected to retard reading. However, no research has proved this hypothesis to be true. An argument in favor of full justification is that readers do report that uneven right margins seem sloppy. This is certainly a judgment to be avoided in technical work at all costs. Since full justification does not retard the reader while it does improve the appearance, full justification is recommended, especially in reports with long letter-width lines in which the spacings are easy to adjust.

Paragraph format

Paragraphs are the essential intermediate organizing technique in writing, so paragraph format is particularly important. Most professionally printed books and journals use either indented paragraphs or paragraphs separated by a blank line. The latter is called block format. These standard professional formats do not seem to be as effective in ordinary technical documents. The indentations are less distinct when used in a single page width column, and many lines are skipped in technical writing just to make room for equations and other exhibits. Also, in student work, it is helpful to emphasize paragraphs because of their important organizing role. Consequently, for ordinary technical documents and especially for student work, it may be advisable to indent *and* to insert a blank line after every paragraph. The extra line helps identify a new paragraph, and the indention avoids confusion with lines and blocks that are skipped for exhibits. This combination may be called the *modified block format*. It is common in technical reports prepared with personal computers and printers, and it is applied in all of the sample reports presented in this text.

Type size

Practitioners and experts seem to agree that any type size between 9 and 12 point is generally acceptable (Tinker, 1963). Consequently, specific requirements should call for a suitable size in this range, probably 10 or 12 point type. If no specific size is required, pick a suitable size in this range and apply it uniformly. A larger type size, such as 12 point, should be used for ordinary, and presumably lower resolution, office printers and copiers. Professional printing should allow for smaller sizes.

Font

The font is the graphical design of the type face. The biggest differences among fonts are the system of spacing and the presence of serifs. Serifs are the tiny lines at the ends of the long strokes. Proportional fonts have variable letter spaces depending on the size needed for the specific letter. **Times New Roman (TNR)** is a proportional serif font, as is Garamond, the font of this text. The **Arial** font used in many graphics is a proportional sans serif (*i.e.,* without serifs) font. `Courier` is a widely available non-proportional or fixed-spaced serif font. Note that the space for an "`i`" in `Courier` is as wide as the space for an "`m`."

Practitioners seem to agree that proportional spacing and serifs tend to draw a group of letters together visually into a unitary word or even a phrase. One would think this design feature would accelerate reading. Unfortunately, research does not necessarily support this idea. Indeed Tinker (1963) studied a number of common type faces of different styles and found that all were equally legible; however, he also reported that readers did not prefer a sans serif type. Since the research is equivocal, it is advisable to follow practice and assume that a proportional serif font is easier to read, at least in lengthy text. Consequently, use a serif font, such as **TNR**, for all extensive text. In contrast, a sans serif font such as **Arial** is thought to slow down the reader's eyes. It is common practice to limit sans serif text to very brief notes such as titles and headings or notes on figures and presentation panels. These are applications where special attention and slower reading is actually desirable.

Some editors will call for a fixed spaced font such as `Courier`. Although readers may expect to see a proportional font, no conclusive research can refute this choice. Courier does seem to be especially helpful when the individual letters—and not the word or phrase—are emphasized, such as when quoting a URL or for numerical data. For example <`www.collegepub.com`> in Courier may be easier to read and enter precisely than <www.collegepub.com> in TNR. In summary, pick one suitable font, or choose one font for the body and one for headings only. Then, apply this choice consistently with occasional exceptions for the special cases mentioned above. Above all, do not use multiple fonts randomly and thereby introduce the highly undesirable *ransom* note effect.

Italic type

Italic type was developed to mimic handwriting. In general text, it has been universally used for foreign words and phrases and for the titles of books and journals when listing references. Probably because it resembles handwriting, it has come to be used in technical work for math symbols in text, tables, and equations. Italic does stand out, but it is definitely slower to read (Tinker, 1963). Therefore, italic should be reserved only for those specific conventional applications.

Paragraph Formats

Most of the problems and the bother of paragraph design and heading formats can be avoided by using suitable paragraph styles in Microsoft Word. A paragraph style is just a preset format that is available from a pull down list in the Formatting toolbar. For example, the style for a "Normal" paragraph should be 10 point TNR, 1.5 line spacing, and full justified. The "Equation" paragraph should be similar with a center tab at the center of the line and a right-flush tab at the right margin. Suitable styles for the prescribed various levels of headings should also be provided. An easy way to include the needed styles is to begin with a starting document that includes these styles. A suitable starting document is available on the web page of one of the authors (Jeter, 2002a).

Units

Units of measurement are a distinctive and absolutely critical feature in technical writing. Most engineering institutions and organizations such as the AIChE, ASCE, ASME, and

IEEE have adopted the Système Internationale (SI) units as the required sole or primary system. Consequently, SI should be used as the primary units for all undergraduate reports. Organizations such as ASCE and ASME also allow or even encourage SI units to be accompanied by United States Conventional Units (USCS). In practice, many industries and even some professional organizations such as ASHRAE still use USCS as the primary units, and they will surely continue to do so. Consequently, the student is well advised to be familiar with both systems, and the technical author addressing the widest possible audience should try to include both systems when appropriate and allowed. For further guidance, probably the handiest and one of the best references on units is the online reference on constants, units, and uncertainty that has been developed and is maintained by the NIST (Taylor and Mohr, 2002). Some specific guidelines are given below.

Primary units

Students should uniformly use SI as the primary system in all reports. In text, accompany the SI values and units with USCS values and units in parentheses when desired and appropriate. An example is 1.05 m (3.44 ft or 41.3 in.). When converting units, do not be tempted to use excessive digits that imply a greater accuracy in the converted data than in the original. For example 1.05 m (41.3 in.) is acceptable, but 1 m (39.37 in.) is not. Note the recommended special case of the period in the abbreviation "in." for inch. This period is unique; otherwise, do not end an abbreviation for a unit with a period. When data in secondary units are desired in tables, use an adjacent column for the accompanying data.

Elementary SI units

The SI is based on seven basic or elementary units of measurement that are summarized in Table 3.1 below. Note that an abbreviation is not capitalized unless it comes from a person's name. Also note that the name of the unit is not capitalized in text even when it does come from the name of a person.

Table 3.1. Elementary Dimensions and Units in the SI

Elementary Dimension	Name	Symbol
length	meter	m
mass	kilogram	kg
time	second	s
electric current	ampere	A
absolute temperature	kelvin	K
molar amount of substance	mole	mol
luminous intensity	candela	cd

Derived units

Numerous derived units have been defined based on the elementary units above. An obvious example is the mass density, which may be written as km/m^3 or $kg\ m^{-3}$ or $kg{\cdot}m^{-3}$. Note that division by a unit is indicated by the solidus, or slash dividing line, or a negative exponent. Multiplication of units is indicated by a blank space or a multiplication dot. Note that this dot is available from the usual symbol set of characters. Twenty-two derived units have been given names. Some of the more common derived units are given in Table 3.2. See the NIST web reference (Taylor and Mohr, 2002) or any equivalent resource for a full list of the derived units.

Table 3.2. Common Combined Dimensions and Derived Units in the SI

Combined Dimension	Name of Unit	Symbol	Elementary Definition	Other Definition
plane angle	radian	rad	$m{\cdot}m^{-1}$	none
solid angle	steradian	sr	$m^2{\cdot}m^{-2}$	none
frequency	hertz	Hz	s^{-1}	none
force	newton	N	$m{\cdot}kg{\cdot}s^{-2}$	none
pressure, stress	pascal	Pa	$m^{-1}{\cdot}kg{\cdot}s^{-2}$	N/m^2
energy, work, quantity of heat	joule	J	$m^2{\cdot}kg{\cdot}s^{-2}$	$N{\cdot}m$
power, radiant flux	watt	W	$m^2{\cdot}kg{\cdot}s^{-3}$	J/s
electric charge	coulomb	C	none	$s{\cdot}A$
electric potential electromotive force	volt	V	$m^2{\cdot}kg{\cdot}s^{-3}{\cdot}A^{-1}$	W/A
capacitance	farad	F	$m^{-2}{\cdot}kg^{-1}{\cdot}s^4{\cdot}A^2$	C/V
electric resistance	ohm	W	$m^2{\cdot}kg{\cdot}s^{-3}{\cdot}A^{-2}$	V/A
electric conductance	siemens	S	$m^{-2}{\cdot}kg^{-1}{\cdot}s^3{\cdot}A^2$	A/V
magnetic flux	weber	Wb	$m^2{\cdot}kg{\cdot}s^{-2}{\cdot}A^{-1}$	$V{\cdot}s$
magnetic flux density	tesla	T	$kg{\cdot}s^{-2}{\cdot}A^{-1}$	Wb/m^2
inductance	henry	H	$m^2{\cdot}kg{\cdot}s^{-2}{\cdot}A^{-2}$	Wb/A
relative temperature*	degree Celsius	°C	$K - 273.15$	none

*Note that the relative temperature requires special attention. Please see the section on this topic.

Additional units

Several units are commonly used along with the SI by convention even though they are not on the recommended list. These include the obvious day, hour, and minute of time. Another obvious addition is the angular degree (*i.e.*, 180° = π rad). A rather important extra unit is the liter (*i.e.*, 1 L = .001 m³). Note that the capital L is the preferred symbol used to avoid confusion with the numeral 1. The metric tonne (*i.e.*, 1 tonne = 1000 kg) is also included. Note the spelling of *tonne* that can be used to minimize confusion with the conventional ton. While on the subject of tons, it should be mentioned that the ton of refrigeration is best identified as the U.S. Refrigeration Ton (*i.e.*, 1 USRT = 12000 BTU/hr).

Relative temperature

The SI does recognize the relative temperature, so the Celsius degree is defined according to the formula,

$$t_\text{C} = T - 273.15$$

where 273.15 K is the conventional ice point. Differences or derivatives in either Celsius degrees or kelvins are identical. For example, a heat capacity could be written as kJ/kg·K or as kJ/kg·°C. Temperature differences are subject to confusion because a 100°C difference could be misinterpreted as a 373 K temperature instead of a 100 K difference. If differences and derivatives are written in kelvins, there is little or no chance of confusion; consequently, this style probably should be preferred. In a numerical example, consider 1.0 kg water, with a heat capacity of 4.2 kJ/kg·K, being heated from 25°C to 30°C. Then the temperature rise would be 5 K, and the heat required would be 21 kJ.

Conventional units

The conventional units are based on common usage and legal definitions in the USA. Most of the conversions between USCS and SI can be derived from the following exact definitions:

pound mass	1 lb$_\text{m}$	=	0.453 592 37 kg
foot	1 ft	=	0.3048 m
absolute temperature	1 R	=	1.8 K
BTU	1 BTU	=	1.055 055 852 62 kJ

To complete the usual set of conventional mechanical units, note that a pound mass weighs exactly one pound force under the standard gravitational acceleration of 9.806 65 m/s² making the lb$_\text{f}$ very nearly 4.448 222 N. To complete the usual set of thermal units, note that the exact definition above makes the BTU very nearly equal to 778.169 ft·lb$_\text{f}$. For more details and values, a rather complete set of unit conversion factors is available at the NIST web site (Taylor and Mohr, 2002).

Unit formats and abbreviations

Abbreviations should follow standard usage as illustrated in the tables presented earlier. Unit abbreviations are not math symbols; therefore, they are always in vertical text, properly called roman, not italics. Note that the single letter "s" not "sec" is preferred. Double line fractions such as $\dfrac{\text{N}}{\text{m}^2}$ are allowed for derived units, but they are conspicuous and awkward to enter. Note how that fraction disrupted the line spacing unnecessarily. Slash or shilling fractions such as N/m^2 are equally proper and simpler and do not disrupt the line spacing the way a double line fraction does. Obviously, slash fractions are much to be preferred. Periods are not used after abbreviations except in the special case of "in." for inch. Some editors omit the ° sign for the Celsius or Fahrenheit degree and simply write 25 C; however, the SI includes the degree Celsius and specifically includes the degree symbol. The SI requires °C probably to avoid confusion with the C for the coulomb of charge. Obviously, not much confusion with the coulomb is likely in mechanical and chemical engineering literature if C is used for the Celsius temperature; however, organizations that have adopted the SI should probably follow the standard strictly and use °C. In any event, the degree symbol is never used in an absolute K or R temperature.

Other Features Specific to Technical Writing
Abbreviations
Typically, abbreviations should be avoided except for the conventional abbreviations for units that are discussed above. A long technical name can be replaced by an acronym formed from the initial letters of the name such as the acronym TNR used above to represent Times New Roman. The acronym should always be identified by a note in parentheses at the first appearance in a short report and probably at the first appearance in every chapter in a longer work. Even acronyms should be minimized, and they should always be identified in the text. Do not make the reader guess what the acronym means. For example, is LSD a dangerous psychedelic drug or the least significant digit?

Identifying equipment
It is helpful to the reader to describe the engineering or experimental equipment generically, not with merely an unfamiliar commercial name or acronym. For critical instruments, follow up with a commercial identification. One style is to place the commercial description in parentheses. This style emphasizes that the commercial description is secondary. For example, an experiment might use a mercury in glass thermometer (Omega, model ASTM-3C). In this text, this style is called "generic (commercial)." Another style is to put the commercial description in a secondary subordinate clause or auxiliary phrase, such as describing a mercury in glass thermometer, specifically an Omega, model ASTM-3C. For convenience, this text refers to the latter style as "generic then commercial." The generic description explains the fundamental character of the instrumentation to the reader. The commercial identification might be helpful in convincing the reader that the instrumentation was adequate, and it would certainly be helpful in reproducing the experiment. The commercial identification may appear technical and well-informed, but it is essentially superficial. Instead, always emphasize a functional and fundamental description that will be meaningful to the typical technical reader.

References

References are crucial to scholarship and credibility. Use a standard method to identify every reference cited in the text. The Harvard method, citing by name and date such as Newton (1686) or (Newton, 1686), is so simple to implement that it should be preferred for student work. The acceptable alternative is to refer by the index number [1] in the list of references. In the reference section, list every cited item using a standard bibliographical format. When the author-date method is used, list the references alphabetically; and when the index number style is used, list in order of appearance. See some examples at the end of this section and more in the section on references. Never stoop to using illiterate citations to the literature such as writing, "our thermo text shows. . . ." To introduce "well known" information, a handy short cut is a phrase such as "standard texts show . . . ." Detailed information on citing and listing references is included in Chapter 2.6 of this text.

Equation formats

Handwritten equations are difficult to read and edit, and they do not seem professional, so equations must be typed into student reports. It is easier and much preferable to use an equation editor rather than the text mode. For consistency in undergraduate reports, always number the equations in sequence beginning with Equation 1. In the text, refer to a specific equation by number with the index number in parentheses. Consider a reference by specific number to be a proper name, so capitalize the name of an equation as in *Equation 1*. The actual equations should be centered on a separate line:

$$F = m\,a \tag{1}$$

Enclose the equation numbers in parentheses and place the equation numbers flush with the right margin.

Equation grammar

Two interpretations of the grammatical status of an equation are possible. First, an equation is an algebraic sentence or clause with its right hand side being the subject, left hand side being the object, and the equal sign—or inequality sign—being the verb. By this interpretation, an equation can be considered to be a clause that is part of a sentence or a separate sentence. Second, an equation is also a specific idea in mathematical form, so an alternative definition is to think of it as a noun or noun phrase representing an idea. Equation 1 above is considered to be a noun phrase at the end of a sentence. The first interpretation considers the equation to be a clause. Therefore, one could write this compound sentence stating that special relativity relates the energy of a particle to its mass, and

$$E = m\,c^2 \tag{2}$$

Since an equation is an algebraic sentence, logic would indicate that it should end with terminal punctuation, such as a period or comma. However, the terminal punctuation after an equation has long been omitted and would now be considered hopelessly obsolete or just strange, so follow the convention and omit any end marks. A particular example is the period that seems to be needed but should be omitted at the end of an equation standing alone as a sentence or even after an equation ending a sentence of text such as this one:

$$dE = \delta Q - \delta W \tag{3}$$

For detailed information on formatting and editing equations see Chapter 2.10, which is dedicated to this topic.

Capitalizing exhibit names

Equations are the most common exhibits, and the practice used herein for referring to a specific "Equation n" as a capitalized proper name is not universal. Indeed, no practice for referring to exhibit names is universal from publisher to publisher or from organization to organization. However, every individual writer needs a consistent practice, and every course needs a standard policy to allow consistent and fair grading. In undergraduate reports, it is reasonable to adopt the policy that a reference to a particular exhibit by its unique identification number is analogous to referring to a person or place by a proper name. Therefore, one should capitalize exhibit names, as in "as shown in Figure 1" or as in "as seen in Table 10." In contrast, do not capitalize the word "figure" in the phrase "as shown in the figure" because this is a reference to a common name.

Lists

Technical writing can be structured, simplified, and condensed by using lists. Two types of lists are used: "vertical" lists in column format, and lists in text style called "run-in" lists. A complex paragraph or section burdened by minutia can often be simplified and enhanced by moving the details to a list. See Chapter 2.13 for guidelines on the construction and editing of lists.

Parentheses

Parentheses are almost sure to disrupt the natural flow of a sentence, so parenthetical comments should be avoided. Unfortunately, the misuse of parentheses is widespread in technical writing. Indeed, they are the "duct tape" of technical writing, and they are warning signs of disorganization. Parentheses cannot be used to evade the rules of grammar or to shortcut the demands of clear writing, so novice writers should avoid this temptation and minimize their use of parentheses. In general, only the conventional uses of parentheses are recommended. Some common allowable examples include setting off citations and punctuating auxiliary explanatory notes and examples. Some transition is needed to identify the purpose of the parenthetical note, so it is good practice to start such notes and examples with an introduction. The conventional abbreviated introductions are both short and effective. So one can efficiently use either "*i.e.*" meaning "that is" to introduce an explanatory note or "*e.g.*" meaning "for example" to introduce an example. When one of these introductions to some information seems inappropriate, it may not be correct to consider the information as being merely parenthetical. Probably then the information is so important that it belongs in a sentence, or it is so trivial that it should be omitted. Some other acceptable and even mandatory uses peculiar to technical writing are listed in Table 3.3.

Table 3.3. Some Acceptable or Mandatory Uses of Parentheses in Technical Writing

Citing literature in (author, date) method	Providing alternative units as in 1.0 m (3.3 ft)
Punctuating indices in a list such as (1) or (a)	Punctuating equation numbers
Punctuating a source note in a table	Identifying a coordinate point as in (x, y)
Introducing an unfamiliar acronym as in laser Doppler anemometer (LDV)	Providing alternative commercial identification as in "dual beam LDV (TSI System 9000)"
Defining an open interval as in (a, b)	In algebraic formulas such as $b\,(a + c)$

Summary Table

Table 3.4 summarizes the guidelines in this chapter. A more detailed checklist is available online (Jeter, 2002b) and may be downloaded and used as a guide to editing, proofreading, and grading text features in undergraduate lab reports.

References

Kolers, P. A., R. L. Duchnicky, and D. C. Furgerson, 1981, "Eye Movement Measurements for Readability of CRT Screens," *Human Factors,* Vol. 23, No. 5, pp. 517–527.

Jeter, S. M., 2002a, "Example Document, EXDOC.doc," ME 4053 Engineering Systems Laboratory, the George W. Woodruff School of Mechanical Engineering, Georgia Institute of Technology, Atlanta, GA, 4 January, document is available online at <me.gatech.edu/~sjeter>.

Jeter, S. M., 2002b, "Generic Guideline and Check List," ME 4053 Engineering Systems Laboratory, the George W. Woodruff School of Mechanical Engineering, Georgia Institute of Technology, Atlanta, GA, 4 January, document is available online at <me.gatech.edu/~sjeter>.

Taylor, B. N. and P. J. Mohr, 2002, "The NIST Reference on Constants, Units, and Uncertainty," NIST Physics Laboratory, National Institute of Standards and Technology, NIST Physics Laboratory, 100 Bureau Drive, Stop 8400, Gaithersburg, MD 20899-8400, reference is available online at <http://physics.nist.gov/cuu/>.

Tinker, M. A., 1963, *Legibility of Text,* Iowa State University Press, Ames, Iowa.

Table 3.4. Guide to Editing Some Features in Text

Use a formal and legible font.	Avoid multiple fonts.
Use standard abbreviations.	Degree sign for temperature differences only.
Distinguish in. for inch.	Follow capitalization rules in abbreviations.
Use generic (commercial) or generic then commercial style of identification.	Avoid superficial identification.
Indent first line in semiblock paragraph.	Indent lists as needed.
SI units must be primary.	USCS may accompany in parentheses.
Sources must have standard citation.	References must have standard listing.
No informal citations.	No incomplete or informal listings.
Equations must be printed.	Use equation editor not text.
Every equation must be numbered.	Center equation on separate line.
Use lists to itemize and organize complex material.	Lists must be rhetorically parallel.
Avoid parenthetical discussions.	Minimize unnecessary parentheses.
Use parentheses when customary.	Use traditional introductions in parentheses (*e.g.*, *i.e.*)
Proofread personally; do not rely only on the spell-checker.	Proofread backward sentence by sentence for grammar.
Proofread backward word by word for spelling and usage.	Avoid common usage errors (*e.g.*, it's for the possessive).

Chapter 2.4

General Guidelines for Editing Exhibits

Exhibits are crucial in technical writing. The exhibits explained here are graphs, illustrations, and other features that accompany the body of the text. One should become familiar with and use the editing guidelines, which are summarized in this chapter and further detailed in following chapters. As with the other guidelines presented here, these recommendations represent the apparent consensus in present-day technical writing.

Cover page

For many reports, especially longer reports, a cover page is appropriate or required. If a cover is used, one should make it professional by following the ISO standard. See Chapter 2.7 for guidelines and an example.

Graphs

Graphs are probably the most important exhibits in any technical report, so they deserve special attention. The primary rule is to avoid clutter or illegible features in any graph. For details, please see Chapter 2.8. Here, we present a few salient comments. Every graph or other type of figure should have a unique number and a descriptive title. By convention, the title is a caption, and it is always located below the graph or illustration. Reserve markers to represent actual experimental data points. Instead of using meaningless markers, use distinct line types to identify model and literature curves. Almost always avoid the Microsoft Excel or Quattro Pro "line graph"; use the "XY" or scatter graph. Consider, understand, and respond to the significance of the $(0, 0)$ point. If this origin is physically significant, as with scaled variables, then one or both of the axes should begin at zero. The graph utility will automatically select (x_{min}, y_{min}) for the origin to maximize the resolution. If this selection should be overridden, you should manually adjust the axes to specify $(0, 0)$ or possibly $(0, y_{min})$ or $(x_{min}, 0)$ as the origin of coordinates. For further guidelines and detailed instructions, please see Chapter 2.8, which is dedicated to graphs.

Illustrations

Other figures, such as schematics of apparatus, are often included in technical reports. Keep these illustrations simple, legible, and uncluttered. For further guidelines and detailed instructions, please see Chapter 2.9, which is dedicated to illustrations.

Equations

These have been discussed as a text element, so please refer to that topic as discussed earlier in Chapter 2.3 for an overview, as well as the detailed guidelines in Chapter 2.10. Microsoft

Word users should ensure that the Equation Editor, actually a minimal form of MathType, has been installed in their computers to allow for the production of high quality equations.

Spreadsheets

Since they are widely used for processing experimental data, spreadsheets are commonly required attachments for most undergraduate lab reports. They are used in monitoring one's data and calculations. Microsoft Excel users should ensure that the Data Analysis package has been installed and identified as a so-called Add-In feature to allow complete regression analysis and to activate all of the spreadsheet functions. A spreadsheet is usually included as an attachment, which is a single item appendix, to a report. Chapter 2.11 presents guidelines on preparing the spreadsheet file to receive and process data. Entire spreadsheets or even large blocks are usually too expansive and informal to be included in widely published professional reports. In contrast, a block from a spreadsheet or even an entire concise spreadsheet may be attached to a report designed for limited circulation. As noted above, the most pertinent data should almost always be extracted and presented in a separate and concise table integrated into the body of the report. Be sure that this attached spreadsheet has a unique number and name that is centered and prominent on the page such as "Attachment 1. Experimental Spreadsheet."

Printing or attaching spreadsheets

A spreadsheet may be printed out separately and appended to a report. Preferably, a picture of a suitable and pertinent block can be attached to a report. When printing out a spreadsheet, pay close attention to the page design. If several pages are required, be sure that each individual page is coherent and meaningful. A printout is sometimes acceptable, but it is rather informal. Creating a concise spreadsheet attachment is the preferred presentation. An attachment can be easily created from an Excel spreadsheet. The easiest method is to point out the desired block and then copy it as a picture into the report document. Detailed instructions are presented in Chapter 2.11. In either case, do not print a page of isolated columns with no apparent purpose. If some calculation in the spreadsheet is important, unusual, or complicated enough to require special description and explanation, it may be appropriate to print out a small block of cell formulas that illustrates the calculation in a separate attachment with a separate identification. Never print a huge block of formulas along with less relevant information such as numerical data, labels, and so forth. Print only a pertinent, representative block implementing the critical calculation or algorithm such as a numerical integration. It is good practice to describe the calculation or algorithm represented by the block of cells in an accompanying paragraph of text.

Tables

Integrating a concise, well-organized table into the body of the report is much preferable to only appending and citing a longer and typically less formal spreadsheet. The unique number and descriptive title of the table should be centered above the block of data. Some editors insist that the unique identifier, *e.g.,* "Table 1," appears on a separate line above the title, but this practice is not universal and is not required in most undergraduate lab courses.

Details in tables

In a table, be sure to label the columns and include the units as needed. Since the primary units are SI, the units in the table should typically be SI. An uncommon exception would be a specific requirement to report raw data measured in conventional or special units. Accompany such data with a column converted to SI. Eliminate insignificant trailing digits. For example, use 3.34 m/sec rather than 3.342 m/sec for a speed measured to about 1% accuracy. Consider using Courier to make the numerals more legible. The following example, Table 1.a, shows minimally acceptable practice, and Table 1.b shows better practice. For more guidance, see Chapter 2.12.

Table 1.a. Velocity Profile Data

position mm	air speed m/s
0.15	1.23
0.2	2.01
0.4	4.52

Table 1.b. Velocity Profile Data from Pitot Probe Scan

position mm	air speed m/s
0.15	1.23
0.2	2.01
0.4	4.52

Lists

Lists have been discussed thoroughly as a text element, so please see that topic as discussed in Chapter 2.3 and the detailed guidelines in Chapter 2.13, which is dedicated to lists.

Appendices

Use appendices or other similar attachments for long mathematical derivations or calculations or for presenting other auxiliary work. Handwritten or manually lettered exhibits or attachments in engineering reports are no longer considered to meet professional standards; therefore, such attachments including their titles must be computer processed and machine printed. Fortunately, state-of-the-art word processors, equation editors, spreadsheet programs, engineering calculation packages, and mathematics packages are now almost universally available. This software is very efficient in processing the necessary calculations and preparing the related documentation. Therefore, use this software to generate or prepare appendices such as sample or auxiliary calculations; and do not submit handwritten work. Such work will definitely not meet professional standards and is unlikely to even be legible. The extra step of preparing a machine processed attachment will also prompt the author to organize and simplify the presentation, probably saving some time for the author and definitely saving much time for the reader. Appendices may be identified by a letter (*e.g.,* Appendix A) or with a Roman numeral (*e.g.,* Appendix I). Give every appendix a unique number and a descriptive title (*e.g.,* Appendix A. Sample Energy Balance Calculations). Every appendix must be cited in the text; for consistency in undergraduate reports, assume that a reference to a particular appendix by its unique identification number is analogous to referring to a person or place by a proper name. Therefore, one should capitalize in "as shown in Appendix 1" or "as detailed in Appendix A" but not in "as shown in the appendix." Remember that appendices are still part of the report and that all features and details must maintain high standards of technical accuracy, scholarship, and production quality.

CHAPTER 2.5

CONSOLIDATED EDITING CHECKLIST

The following checklist is largely consolidated from the pertinent chapters of this text. Refer to the subject section for further guidance. This checklist is a general and generic guideline only. Students are responsible for a neat and professional presentation that is legible and cogent. Every possible error cannot be listed here, but many common problems are addressed.

Table 5.1. Consolidated Editing Guideline

General Aspects

Report is prepared in correct format, *e.g.,* itemized or memorandum.	Cover page, abstract, and table of contents included if required.
Report is dated and signed.	All required or necessary items and topics are addressed.
The writing style is effective, concise, scientific, and efficient.	The report is complete and well organized.
Padded text and redundant comments are avoided with no excessive length.	Jargon or colloquial writing avoided.
Conjectures and vague or subjective arguments are avoided.	All conclusions are justified on scientific and empirical basis.
Text is grammatical and proofread, not merely spell-checked.	The report is securely and safely bound.
Introduction defines the project and answers the questions, WWWWH and W?*	Closure is complete, concise, and effective.
References	
Citing and listing of references is conventional and scholarly.	Page and text formats are conventional, correct, and consistent.
Exhibits and Attachments	
All exhibits and attachments cited are included.	All exhibits and attachments included are cited.
All exhibits and attachments are of professional quality.	Named exhibits (*e.g.,* "Table 1" not "the table") are capitalized.

*Who, what, when, where, how, and why?

Table 5.1. Consolidated Editing Guideline *(cont'd)*

Units and Digits Displayed	
SI units are primary (USCS may accompany).	No insignificant digits in text or tables.
Statistics	
Adequate attention is given any statistics and regression analysis.	R-squared and alpha risk addressed for every regression model.
Page, Section, and Paragraph Formats	
Margins are uniform on top and bottom and on both sides, usually one inch.	No wide bottom margins appear, especially at exhibits.
Section headings are used and accurately identify the role of the text.	Text in a section addresses the topic completely without misplaced remarks.
Modified block paragraph or other required format used consistently.	Paragraphs throughout text have clear objectives and address requirements.
Paragraph Style	
Paragraph topic sentence is present.	Paragraph topic is developed.
Extra line is skipped between paragraphs.	
Sentence Style	
A half-line is skipped between every line of text.	Conventional sentence grammar is used.
Incomplete fragments or run-on sentences are avoided.	Simple sentences predominate with only isolated compound or complex sentences.
The text is impersonal.	First person is avoided or minimized.
Active voice used judiciously.	Parallel structure in list or series (*e.g.,* imperative sentences in listed procedure).
Common typo errors (*e.g.,* "too" for "to," "it's" for "its") are avoided.	Common technical writing faults are avoided.

Table 5.1. Consolidated Editing Guideline *(cont'd)*

Parentheses in Technical Writing

Citing literature in (author, date) method.	Providing alternative units as in 1.0 m (3.3 ft).
Punctuating indices in a list such as (1) or (a).	Punctuating equation numbers.
Punctuating a source note in a table.	Identifying a coordinate point as in (x, y).
Introducing an unfamiliar acronym, as in laser Doppler anemometer (LDV).	Providing alternative commercial identification as in "dual beam LDV (TSI System 9000)."
Defining an open interval as in (a, b).	In algebraic formulas such as $b(a + c)$.

Features in Text

Use a formal and legible font.	Avoid multiple fonts.
Use standard abbreviations.	Degree sign for Celsius and Fahrenheit temperatures only.
Distinguish in. for inch.	Follow capitalization rules in abbreviations.
Use generic (commercial) or "generic then commercial" style of identification.	Avoid superficial identification.
Indent first line in semiblock paragraph.	Indent lists as needed.
SI units must be primary.	USCS may accompany in parentheses.
Sources must have standard citation.	References must have standard listing.
No informal citations.	No incomplete or informal listings.
Equations must be printed.	Use equation editor, not text.
Every equation must be numbered.	Center equation on separate line.
Use lists to itemize and organize complex material.	List must be rhetorically parallel.
Avoid parenthetical discussions.	Minimize unnecessary parentheses.
Use parentheses when customary.	Use traditional introductions in parentheses (*e.g., i.e.*)
Proofread personally; don't rely only on spell checker.	Proofread backwards sentence by sentence for grammar.
Proofread backwards word by word for spelling and usage.	Avoid common usage errors (*e.g.,* "it's" for the possessive).

135

Table 5.1. Consolidated Editing Guideline *(cont'd)*

Citing and Listing References

Recognized citation style is used, usually the author-date style.	One citation style is used throughout except for unusual circumstance of occasionally citing numerous works.
All sources cited are listed in the reference section.	All references listed in the reference section are cited in the text.
Reference listing has complete bibliographical information.	Consistent capitalization system used in all listings of references.
Listing is in proper order with proper punctuation.	Italics are used for books and journal titles, and quotes are used around titles of reports and other shorter works.

Graphs for Experimental Engineering Reports

Design is neat, uncluttered, legible, and effective.	A unique number is assigned and a descriptive caption is given.
Caption always below graph in written report.	A consistent capitalization system is used in all captions.
XY graph, not Line chart, is almost always used.	Graph box included only with discretion and then consistently used in one report.
No reliance on color or subtle gray scales.	No shading of graph area.
Graph is cited in text.	Any cited graph is included.
Graph is inserted into text at first feasible location when possible or on separate page when necessary.	Graph is attached (*i.e.*, as a one item appendix) to extended abstract only.
Full page graph on separate page in landscape orientation has caption to right.	Separately printed graph is pasted onto page or has machine printed caption pasted on to graph page.
Zero point is purposefully and appropriately included or excluded in axes.	Secondary axis is included when necessary.
Axes are not so compressed or span is not so great that variation in data is not apparent.	Special care is taken to not obscure scaling relations between normalized variables.
Axes are titled, with dimensions if existing.	Markers are strictly reserved for discrete data.
Data is presented in scatter ploy or connected profile of markers as appropriate.	Distinct line types (*i.e.*, not markers) are used to identify continuous data or model curves.
The series in a single series graph is identified only in caption.	All series are identified in legend, in notes on graph, or in caption.

Table 5.1. Consolidated Editing Guideline *(cont'd)*

Illustrations and Drawings

Figure is neat, uncluttered, legible, and effective.	Unique number is assigned and descriptive caption is given.
Caption is always below figure in written report.	Consistent capitalization system used in all captions.
Photograph used only with discretion and usually accompanied by schematic.	Usually a schematic or simplified drawing is preferred.

Equations

Equation has logical grammatical setting.	Equation is neat, uncluttered, legible, and correct.
Equation is generated with Equation Editor, not a line of text.	Equation is centered on separate line with unique number at right margin.
Equation is proper size, especially for projection slide.	Equation box is sufficiently wide to avoid truncation.
Templates from menu are used for parentheses, not symbols typed from keyboard.	Math symbols are in italics with functions, numbers, and notes in vertical roman type.

Spreadsheets

Overall design and detailed format are neat, uncluttered, legible, and effective.	A unique number is assigned and a descriptive title is given in heading.
Spreadsheet is presented as an attachment cited in text.	Concise block of pertinent data is also extracted and presented in a table.
Title heading is centered and prominent with a type size of at least 10 point.	Spreadsheet is structured with heading, preliminaries, summary, and data blocks.
Heading identifies purpose and history of spreadsheet.	Unique data, constants, and parameters are placed in the preliminary section.
Summary section is near top of spreadsheet.	Routine recurring data are in a well-organized separate section.
Consistent numerical format is used in a related block or blocks of data.	Units are identified for every dimensional data block, typically in column heading.

Table 5.1. Consolidated Editing Guideline *(cont'd)*

Spreadsheets *(cont'd)*

Recurring data are identified, usually with column header.	Unique data are identified, usually with identifying phrase.
Printed margins are adequate and at least 1 inch all around.	Multiple page printout is carefully designed especially in continuing pages.
Block of formulas presented in separate attachment if required.	Optional concise summary regression block prepared if appropriate.

Tables

Table is cited in text.	Table is concise, neat, uncluttered, legible, and internally consistent.
Table has unique number and descriptive title in heading.	Table is centered horizontally on the page.
No insignificant digits are displayed.	Data are identified, usually in column headers, and units are given unless data are nondimensional.
Table is limited to one page.	Numbers are properly aligned.
Consistent numerical display format is used, except as limited by significant digits.	Scientific notation is used for very large or very small magnitudes or to carefully identify significant digits.
Appropriate text format, usually centered, is used for headers and notes.	Source and descriptive notes and footnotes are provided.

Lists

List is used when appropriate and helpful to simplify and enhance presentation.	List is not used for incommensurate items.
Run-in list has no extraneous punctuation.	Vertical list can have bullets, numbers, letters, or no indices, as appropriate.
Multiple column list is used for brief items.	Optional right indention of vertical list is helpful.
List must be rhetorically parallel.	Consistent capitalization system used in all lists, usually initial capital only.

CHAPTER 2.6

CITING AND LISTING REFERENCES

Citations

Citations are the notes in the text that identify a source of information from the technical literature or other resource. The references are the listing of the sources. One should cite all references in the text and list the bibliographical references at the end of the report. Two citation styles are popular and acceptable. You may cite by number or by the author and date of publication. Usually, citation styles should not be mixed, but in one document, either is acceptable. For student work, the instructor will probably prescribe one style for consistency. Since the author-date method is the easiest to manage, it is a good choice for student and most professional work. The method developed by the Modern Language Association (MLA) from the liberal arts is efficient, but it is never used in engineering. Citation by footnote is widely practiced in legal studies and the humanities, but this method is essentially obsolete in science and technology.

The numerical format cites a reference by its sequential list number. The number is usually typed in brackets, *e.g.,* [1] or Reference [1]. At the end of a sentence, the closing bracket precedes the period. Numbered references should be listed in order of appearance, with alphabetical order as a barely acceptable alternative. When a text that is under development is altered to insert a numbered reference, the index on every succeeding reference must be altered. This extra work can be a nuisance. An automatic sequence number, such as the field code called "Seq" in MS Word, can be used, but the entries and the needed cross references can be awkward. The inserted reference will also cause the list of references to require updating. Obviously, citation by number is effective but somewhat difficult to manage.

The author-date method is also called the "Harvard format." It identifies the source with the author's last name and the year of publication, *e.g.,* (Smith, 1989) or Smith (1989). This method is probably more convenient for undergraduate reports. As just exemplified, the name of the author may be part of the sentence, or it may be parenthetical, depending on the context. To distinguish publications from the same year, use a letter such as Smith (1989a) and Smith (1989b) in the citation and listing. For publications with two authors use (Smith and Jones, 1989) or Smith and Jones (1989). For works by multiple authors, use (Smith, *et al.*, 1989) or Smith, *et al.* (1989). References cited using the author-date format should be listed in alphabetical order. If a citation is added to a report being written, it is easy to just insert the new reference in alphabetical order in the listing. In addition, including the page number in a citation by author-date is usually allowed. An example could be White (1974, pp. 74–76). Otherwise, the page number must be included in the reference listing, which is an awkward and less flexible method. The simplicity and adaptability of the author-date method makes it ideal for student work.

It is sometimes, but rarely, appropriate to mix numerical with author-date citations. A possible example is the specific case in which many references must be cited at once. An example is: "Smith (1989) and References [1], [5], and [18] through [29] show that. . . ." Otherwise, select one style, probably the author-date style, and use it exclusively.

Pay close attention to how sources are cited and listed; appropriate treatment of references is crucial to good technical scholarship. All the cited sources should be listed in the reference section. Acceptable formats for listings are reviewed next.

Listings

A complete list of the bibliographical references cited in the text, with every entry in an approved format, should follow the main text of the report. Please note that various publishers and organizations will differ in style and content of the listings. Nevertheless, your listing must be descriptive, definitive, and consistent. An adequately descriptive listing identifies the source and the nature of the source. The reader needs to know not only the author, but the subject and the publisher of the reference as well. A knowledgeable reader will appreciate recognized authorities and distinguish refereed from less reliable nonrefereed or commercial literature. A definitive listing is complete enough that the reader could actually find the source independently. Finally, the listings should be consistent both among themselves and with the guidelines of the editor or, in academe, the instructor. Remember, incomplete or inconsistent listings draw unwanted attention and raise questions about the writer's attention to content and details. Always record a complete, not minimal, description of your sources during your literature research so that you can generate later whatever listing is required.

Unless some specific contrary rules are given, the student or engineer can safely use the following styles listed here. The styles presented here are a consensus compromise of methods, recommended by ASME, ASHRAE, the University of Chicago Press (1993), and others. The latter is the best resource if you encounter, and cannot circumvent, an unusual problem. Even if another style is eventually required, the student is safe to begin work with the system described here in mind. Since this system is complete and consistent, references that fit it can easily be altered to satisfy any reasonable alternative rules. There is only minimal consistency among editors about details such as the order of the parts and the separators (*i.e.*, commas or periods). Consequently, always understand and follow the particular style that may be required. The list itself may be typed single spaced with a blank line between each entry. There is even a difference in the style of indenting the list. Some editors indent the first line of an entry while others print the first line flush and indent the subsequent lines. The recommendation here is to keep the list simple. Merely indent the first line like an ordinary paragraph, and number the entries only if citing by number. Some general forms and several specific examples for various types of literature sources follow. Some editors allow the initials to be written first in the names of the second and any subsequent authors (*e.g.*, Watson, J. D. and F. Crick, 1953 . . .). In some of the following examples, this highly desirable alternative simplifying detail is used.

Journal articles

Journals are periodical technical or professional publications such as the ASME journals in heat transfer, fluid mechanics, and other fields. Journals usually publish the results of significant and original research. Usually most technical references are to journals, and the listing of a journal reference is about the most typical type. The general form is as follows:

> Optional Serial Number. Last name of first author, followed by initials, Last name of any other authors, followed by initials, Year of publication unless always citing by number, "Title of the Specific Article in Quotes," *Name of the Journal in Italic Type*, Volume No., Issue No., (if citing by number, optional day and month and required year), page range of article.

A typical example follows:

> Wepfer, W. J., and Oehmke, R. L. T., 1985, "Computers in the Mechanical Engineering Instrumentation Laboratory at Georgia Tech," *Intl. J. Appl. Engng. Ed.*, Vol. 1, No. 6, pp. 415–421.

Alternatively, when citing by number, write the initials first in the second author's name, and employ the alternative style of indenting the entry as follows:

> 1. Wepfer, W. J. and R. L. T. Oehmke, "Computers in the Mechanical Engineering Instrumentation Laboratory at Georgia Tech," *Intl. J. Appl. Engng. Ed.*, Vol. 1, No. 6, June 1975, pp. 415–421.

The latter text format is the classic "bibliographical" style characterized by the "hanging indentation" in the second and any following line.

Articles in transactions or conference proceedings

These compilations are the published records of the papers presented at the regular meetings of a technical society or at a special technical conference. This type is nearly identical to a journal publication except the title of the transaction or proceeding replaces the title of the journal as follows:

> Lyons, D. P., Wepfer, W. J., and Oehmke, R. L. T., 1987, "Spreadsheet Analysis of Laboratory Data," *ASHRAE Transactions*, Vol. 93, Pt. 1, pp. 417–424.

Books

Books are longer works usually edited and distributed by a publisher. The publisher is assumed to have some independence from the author and some supervision of the contents. Technical books usually are only secondary references based on primary references that are usually journal publications. The best practice is to review and cite the primary reference. Always distinguish books from long reports, which have no independent publisher. Research reports are discussed next. The listing for a book is actually slightly simpler than a journal since no article title must usually be included. When necessary, identify the page range where the pertinent information is located. Always identify the publisher and the address of the publisher as follows:

> Last name of first author, followed by initials, Last name of any other authors, followed by initials, Year of publication, *Title of the Book in Italic Type*, Edition No. if necessary, Publisher, City address of publisher, sometimes omitted if New York (if citing by number, year of publication), pertinent page range.

A typical example for a book follows:

> White, F. M., 1974, *Viscous Fluid Flow*, 2nd ed., McGraw-Hill, New York, pp. 74–76.

Obviously, the edition number is not needed on a first edition unless a second or later edition is known to exist and would not be a source or would change a part of the citation such as the page range.

Research reports with limited circulation

This category includes all internal reports and reports with limited and/or restricted circulation. Short reports may resemble journal articles, and longer reports may physically resemble books. Nevertheless, avoid assigning them the same status since the author of a report is essentially the publisher and no independent review or supervision can be assumed. This listing is similar to that for a book, except the title is placed in quotes instead of being printed in italics. In addition, include any identifying number for the report and identify the sponsor if known and appropriate. Include the institutional or corporate affiliation of the authors, the entity that in effect serves as the publisher of the report. An example of interest to a small group of undergraduate lab students follows:

> Pascual, C. C. and S. M. Jeter, 1998, "Measurement of Heat Leak from the Copper Cylinder Used in Convection Heat Transfer Experiment," The George W. Woodruff School of Mechanical Engineering, Georgia Institute of Technology, Atlanta, GA, 23 November 1998.

This general form for a research report can be modified as necessary and used for a range of narrowly distributed writings such as calibration reports, student papers, student theses, and course manuals and materials.

Commercial literature

In engineering and especially in experimental engineering and design, references to commercial literature such as specifications, catalogues, and operating manuals are common. A modified form of the report style should suffice. Look for a copyright note to find a publication date. Assume a corporate author unless a personal author happens to be identified, as in this example:

> Omega, 1995, "The Data Acquisition Systems Handbook," Omega Engineering, Inc., Stamford, CT, pp. Z-8 to Z-9.

Personal communications

Because personal communications such as a conversation or message usually cannot be documented in print, using them as sources should be avoided. Nevertheless, situations arise where such sources must be used, and citations will be necessary. Examples include highly specialized data, an eyewitness account, or a personal observation. The engineering investigation of an accident or failure is one example when such sources may be necessary. Be sure to record any such source with an entry in your research notebook. Use the following general format, always identifying the type of communication, such as letter, lecture, conversation, phone conversation, e-mail, and so forth.

> Last name of correspondent or communicator, followed by initials, year, type of communication, such as letter, lecture, e-mail message, phone conversation, or private conversation, place conducted or initiated, specific date.

An example that might refer to specific information needed in an undergraduate report follows:

> Jeter, S. M., 1998, private conversation, Atlanta, GA, 12 August.

Film or video

Occasionally a motion picture presentation must be cited. Identify any principal performer or presenter as in a technical lecture. Use the following form and identify the technical format, such as 16 or 32 mm film, VHS cassette, or DVD, when possible.

> Title in quotes or Last name of principal if any, followed by initials, year, Title in quotes (when a principal presenter is identified first, format, duration, publisher, type of medium.

An example of interest to undergrad lab students follows:

> Eisenberg, P., 1968, "Cavitation," 16 mm converted to VHS, mins. 20 to 22 of 32 mins. overall, produced by Educational Services, Inc., supervised by the National Committee for Fluid Mechanics Films, sponsored by the National Science Foundation and the Office of Naval Research, VHS video cassette.

Internet postings

More and more students wish to cite WWW-based documents in their scholarly and scientific research. This practice seems very dubious for several reasons. One major reason is that the web is essentially a mass of self-published documents with little review or quality control. Even if this problem is discounted, another problem is the ephemeral nature of the web resources. An author can always revise a document and change its content without necessarily changing the date tag in the HTML code. To exemplify the problem, suppose that a practicing engineer uses data from a web page that turns out to be erroneous. Next, suppose that the author of the web page finds the error and revises the web page but does not insert the new date. Suppose now that the erroneous data causes an engineering failure or scientific error. The engineer cannot justify the use of the original data because the web page has been changed while it still contains the old date. The error now looks like mere carelessness by the engineer rather than good-faith use of published data.

In addition, note Lynch's recent warning (1997) on the practical difficulty of finding web articles:

> The Web, moreover, still lacks standards that would facilitate automated indexing. As a result, documents on the Web are not structured so that programs can reliably extract the routine information that a human indexer might find through a cursory inspection: author, date of publication, length of text, and subject matter. (This information is known as metadata.) A Web crawler might turn up the desired article authored by Jane Doe. But it might also find thousands of other articles in which such a common name is mentioned in the text or in a bibliographic reference.

(As an aside, note that the parentheses are in the original, even though the note is hardly parenthetical.) This note is a caution that an apparently complete reference listing may be hard to track down on the web. Probably everyone has had this experience. When possible, then, include the complete Uniform Resource Locator (URL) when necessary to cite a web page.

An Internet posting is about the least reliable literature source since no review, editorial supervision, or even corporate responsibility can be assumed. Nevertheless, citations to Internet postings are becoming more common even though standard works on report writing are too old to even mention the possibility of such a reference. As shown in the example, the format of a short report can be modified to fit web pages. Assigning a date may be problematic, but try to include the date first posted and the date last updated if available. Always include the complete Uniform Resource Locator (URL) as in this example. In this example, the serif evenly spaced font called Courier is used for the URL because it is a readily available font with distinctive and legible characters for all the numerals and letters.

> Pascual, C. C., 1998, "ME 4054: Thermal Sciences Laboratory," Internet Posting at `http://sk19.me.gatech.edu/labweb/me4054/me4054.htm`, updated 8 January 1998, The George W. Woodruff School of Mechanical Engineering, Georgia Institute of Technology, Atlanta, GA, pp. 1–2.

Because of the inherent unreliability of Internet postings, a researcher would presently be well advised to make formal reference only to works reviewed and edited by traditional standards and either published elsewhere in print or published by one of the emerging Internet journals.

Summary Table

Table 6.1 summarizes the guidelines on citing and listing references in this section and may be used as a guide to grading citations and references in undergraduate lab reports.

Table 6.1. Guide to Citing and Listing References in Experimental Engineering Reports

Recognized citation style is used, usually the author-date style.	One citation style is used throughout except for unusual circumstance of occasionally citing numerous works.
All sources cited are listed in the reference section.	All references listed in the reference section are cited in the text.
Reference listing has complete bibliographical information.	Consistent capitalization system used in all listings of references.
Listing is in proper order with proper punctuation.	Italics are used for books and journal titles and quotes used around titles of reports and other shorter works.

References

Lynch, Clifford, 1997, "Searching the Internet," *Scientific American* (Online), March, 1997. Available online: http://www.sciam.com/0397issue/0397lynch.html

University of Chicago Press, 1993, *The Chicago Manual of Style*, University of Chicago Press, Chicago, IL.

Chapter 2.7

Preparing a Cover Sheet

Any long report and any report submitted to a library or document depository should include a cover sheet or title page. This page should include all the relevant bibliographical information so that a technical librarian can catalog and deposit the report without being forced to peruse the entire document.

According to the ISO (1982), the title page should include the following information in the following order:

1. Any distribution limitation or security classification as described below.
2. Any report identifier such as a serial number issued by a particular organization or an ISBN number for a document published as a book.
3. The name and address of the responsible organization, which is essentially the publisher of the document.
4. A descriptive title, which should be concise but definitive. Include any identification of a specific version (*e.g.*, final or interim) and any revision or edition number.
5. The names of personal authors, using the full name of the author in natural order. The ISO recommends that some emphasis be added to identify the family name or other name by which the author should be addressed professionally. For example Drs. Burdell and Perez might use:

George P. <u>Burdell</u>

Jorge <u>Perez</u> de Burdelo

6. Omit any author block when the corporate author is the publishing organization.
7. The date of publication—meaning when the report is offered to the public or its intended recipient. To avoid any ambiguity, use the European convention on order and spell out or abbreviate the name of the month, as in 29 July 2002 or July 2002.
8. Date of priority in parentheses when appropriate as in articles submitted to technical journals. This is the date when the document or manuscript became essentially complete. It is included to help assert a possible right to the priority of discovery.
9. Any special notes such as approvals, disclaimers, and releases.
10. Repeat of distribution limitation.

Distribution limitation

A report may have a distribution limitation to restrict circulation or duplication to protect commercial trade secrets, patent disclosures, or other intellectual property rights. An example is a designation of "Proprietary Information" with accompanying limitations on use. Use this designation to inhibit someone who receives the report in the course of business from using or handling the information improperly, such as by adopting the information for his own use or by passing the information on to your competitors. Another typical limitation may be for the material to be maintained confidential and to be used only for a particular purpose such as evaluating a proposal. When national security issues are involved, a legally enforceable security designation may be imposed, such as "confidential," "secret," or "top secret." Proper handling of such a document is strictly prescribed by law and official regulations.

An example cover page that should be appropriate in an undergraduate course follows.

Reference

ISO, International Organization for Standardization, 1982, "Documentation—Presentation of scientific and technical reports," International Organization for Standardization, Geneva, 15 March 1985.

S<small>AMPLE</small> C<small>OVER</small> S<small>HEET</small>

PROPRIETARY INFORMATION:
FOR INSTRUCTIONAL USE ONLY
NOT FOR DISSEMINATION OR DUPLICATION

REPORT No. ME4055-2002-A4

Energy Release Evaluation of Cold Fusion Using Adiabatic Bomb Calorimeter
Final Report

Generic Institute of Technology
School of Mechanical Engineering
Experimental Engineering Laboratory Course
Summer Quarter 2002

Group A-4:

George P. <u>Burdell</u>
Georgia <u>Prandtl</u>-Burdel
Georges-Pierre <u>Bourrel</u>
Jorge <u>Perez</u> de Burdelo
Bu <u>Tien</u>

27 July 1999

(Manuscript Completed 21 July 1999)

Instructor Acceptance: _____ Date: _____ Final Grade: _____

CHAPTER **2.8**

GUIDELINES FOR GRAPHS

Introduction

Graphs are probably the most important exhibits in the typical experimental engineering report. The graphs present the data in comprehensive and condensed visual format and reveal tendencies in the data and relations between independent and dependent variables. Since graphs are especially important, their preparation requires special care. Computer-generated graphs are expected in professional work and are usually required in undergraduate labs, except for in the very few instances where graphs printed by stand-alone instruments must be used. Spreadsheet generated graphs, such as graphs from Excel or Quattro-Pro, are strongly preferred. Other software may be used, but they are not recommended. A graph must be legible and uncluttered, and its content must be clearly presented. Some especially important general tasks in designing and using a graph are as follows:

- Citing and identifying the graph
- Specifying the axes and scales
- Identifying the series
- Formatting the overall graph area and the smaller plot area itself

Since Excel is widely used, Excel graphs dominate student work. Probably the biggest time saver in generating acceptable Excel graphs is setting a convenient default graph. Instructions on setting the default graph are detailed later in this chapter. Other specific guidelines and examples follow in the next sections.

Specific Guidelines and Techniques
Graph type
Virtually all Excel or Quattro graphs should be so-called scatter or XY graphs not line charts. The line chart is a business-oriented presentation with all points in a series evenly spaced along the horizontal axis without regard for the x value. In the XY graph, the horizontal position is proportional to the x value, the usual convention for a mathematical graph.

Graph caption
Always caption your graphs in reports with a unique consecutive figure number and a descriptive title. The caption is always placed below the figure. Be sure to cite every figure in the text. This rule applies to every exhibit or attachment. When an exhibit is cited by its specific number, its identification is usually considered to have become a proper name, such as "Figure 1.,"

and as such the name should be capitalized. Along with the unique identifier, use a descriptive title, such as "Figure 1. Velocity and Turbulence from Thermal Anemometer Traverse" exclusively. Avoid a generic name or a mere restatement of the titles of the axes, such as "Figure 1. Velocity vs. Radius." You may elect to print the title above the graph when it is embedded in a spreadsheet page or when it will be presented as a transparency to be projected for an oral and visual presentation; however, most real-world editors demand figures in written reports to have their titles below them, in the form of a caption. Using the caption to identify the graph is easier because the figure number does not need to be known when the graph is first generated. This practice also makes updating the figure number or the text of the caption much easier. Capitalization in the caption should be consistent throughout a report. Either capitalize only the first word or capitalize every main word uniformly. In summary, the recommended method is to insert the graph, as a picture, into the report without any title and to print the caption below the graph as part of the text of the report.

Lines and Markers

Usually, experimental data in mechanical engineering is actually a set of isolated data points. Other data, such as the literature model in Figure 8.1 or a regression model developed from experimental data, are usually represented as a continuous curve. Sometimes, literature data is presented in a table, such as a table of thermodynamic properties, but the table almost always represents merely a set of values computed at arbitrary points from a model, not a set of measured experimental data. Remember these differences and respect the ultimate product when selecting lines or markers.

SAMPLE FIGURE

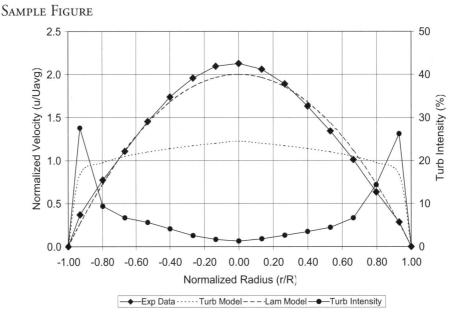

Figure 8.1. Velocity and Turbulence from an LDV Traverse

As shown in Figure 8.1, it is best to reserve markers for experimental data, such as the velocity data in that figure. Do not use markers merely to identify a line. Instead, use a distinct line type to identify a continuous curve, such as the model profiles from the literature in Figure 8.1 or for the regression model based on the data in Figure 8.2. The models may be obtained from an equation or based on a table of values. Even if the model is based on a table of calculated values, illustrate it with a continuous line. Otherwise, these markers can be confused with actual data points. In the unlikely event that a table of experimental data points from the literature is plotted, use markers for these data. On occasion, the experimental data will be a continuous, or nearly continuous, set of points, such as the time trace of an accelerometer. Such an experimental curve can be presented without markers since every point in the curve should be understood as representing an experimental datum. Otherwise reserve markers for data. Depending on the context, it is sometimes appropriate—and sometimes not appropriate—to connect the markers by straight lines as discussed in the next paragraph.

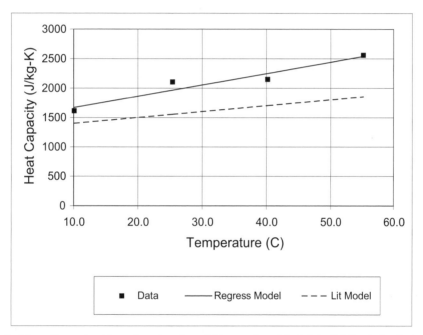

Figure 8.2. Heat Capacity Data (markers) and Regression Model (solid line) for Lubrication Oil Compared with Model from Tempco (1992, broken line)

Profiles or scatter plots

As illustrated in Figure 8.1, sometimes experimental data points should be connected by straight lines in a profile. As shown in the example, such connecting lines are mere tie lines and should be straight, so the "smoothed line" option in Excel should be deselected. In other cases, it is logical to display the data points as a scatter plot of isolated markers, illustrated in Figure 8.2. Specifically, data used to develop a regression model should be plotted as a scatter plot of unconnected markers with the regression model displayed as a solid line without markers.

When feasible, an error band or uncertainty envelope for the regression model should be plotted along with the regression line itself. Use one distinct line type, such as the dashed line, to indicate the upper and lower limits of the error band, as in Figure 8.3, which is presented later as a comparative example. As mentioned earlier, a comparison curve representing either a theoretical model or a regression result from the literature should not have markers; an alternative and distinctive line type is preferred.

Smooth lines or straight lines

The knowledgeable reader knows that no theoretical model or even a reasonable statistical model will pass exactly through experimental data. Consequently, lines used to merely connect markers into a profile, such as for the velocity and turbulence intensity in Figure 8.1, should be straight lines. This choice emphasizes that the lines are merely a convenient graphical way of grouping the points into a distinct set. A theoretical or statistical model is essentially a continuous set, and the appearance of the line drawn to represent this set may be improved by using the "smoothed line" property in Excel.

Identifying series in a graph

At least three methods are used to identify series. The simplest and most direct way is to identify the series in the legend. Recall that, in Excel, the legend label is not accessible as a property of the data series. Instead, the legend label is accessed by the Series tab in the Source Data dialog box. Access the Source Data dialog box from the Chart selection on the Main Menu bar. As always, avoid using markers merely to help identify a line. Such extraneous markers are too easily confused with data, so use distinct line types. An alternative or complementary method is to use labels on the graph plane the way the models are identified in Figure 8.1. Here, the notes identifying the comparative models were added by inserting Word text boxes, an option on the Drawing Toolbar. A final method that may be suitable for a very simple graph is to identify the series in the caption. This method is recommended and even required when only one series is plotted. If only two or three series are plotted, notes in the caption (*e.g.*, parenthetical notes) as with Figure 8.2 can be used. Here, the figure caption is part of the Word document and was placed below the graph as required for technical reports.

Use of colors

Do not rely on colors; rather, use distinct line types or markers with distinct shapes to identify series. Remember that color printers are still rare, and that reports will most likely be printed and duplicated only in black, white, and gray. It is best to use only 100% black (*i.e.,* no pastels that print as indistinct gray scales) for all line and markers. Color can be used in addition for a projection transparency, but ensure that the graph is still legible when printed in black and white.

Axes

Label and select the range for the vertical y and horizontal x axes carefully. A wide range may compress the data excessively. A compressed range may obscure an overall pattern by exaggerating minimal erratic variations. Offset the graph, omitting the zero point on either or both axes, only with care. This omission can distort proportions and obscure scaling relations such as the pump and fan laws. Consider, understand, and respond to the significance of the (0, 0) point. If it is physically significant, such as with many normalized variables, manually adjust the axes to specify zero at the origin of an axis.

Secondary vertical axis

Use a secondary axis to display a series with distinct units or when the magnitude of one series differs markedly from the others, as in Figure 8.1; however, do not allow two confusing sets of gridlines to appear. If two sets of gridlines are used, adjust the specs so that they are collinear. For example, in Figure 8.1, the gridline for the normalized velocity of 1.5 is collinear with the gridline for 30% turbulence intensity. An advantage of normalized data (*e.g.*, u/U or $C_p = P/q$) is that several series can be displayed on one axis without regard to units so long as the magnitudes do not differ too much. When two y axes are used, give each axis a distinct label and include the units for dimensional data.

Graph box

Whether to surround the graph with a text box as in Figure 8.2 or to eliminate this feature is a matter of personal preference if no specific guidance has been given. Avoiding the box eliminates one more feature to worry about and makes it easier to resize graphs without creating ragged right and left margins. Be consistent at least within one report on whether to include or eliminate the graph box. To eliminate the box in Excel, click on the Chart Area (*i.e.*, anywhere outside the graph itself, which is called the Plot Area) and right click to format. Pick "None" for the border option to eliminate the graph box.

Incorporating the graph into the report

It is awkward and only marginally acceptable to print graphs separately and append them to a written report. It is easy to produce a more professional result by integrating graphs into the text rather than attaching a separately printed graph. Separately printed graphs may be attached to an extended abstract used to document an oral and visual presentation. Otherwise, in all undergraduate lab reports, integrate all graphs generated by spreadsheets or other graphics software into the body of the text at the first feasible location following its citation in the text. A graph of typical size, such as the examples in this section, is best integrated into the text in the usual portrait orientation with the caption below the graph. Review the page format before printing to avoid leaving a misleading gap before or after a graph. Reposition or split a paragraph if necessary to avoid the appearance that the end of a major section has been reached. A very complex graph may need to be printed on a separate page. When the graph on a separate page is best presented in landscape orientation (*i.e.*, width greater than height), the caption should be to the right of the reader.

Techniques for inserting graphs into text

Inserting the graph into the text as a picture is handy. This method allows the caption to be printed beneath the figure in the same font used in the balance of the text, and it allows the figure numbers and captions to be readily updated. Note that the graph must be completely edited in the spreadsheet before you copy it as a picture. Avoid lines with fine detail and subdued colors that may not translate well. It is safest to change all of the important graph elements such as lines and markers to 100% black. Minor elements, such as grid lines, can be gray. In either current version of Excel or Quattro-Pro, a good image of a graph may be inserted by copying the graph as a picture from the spreadsheet to the Windows clipboard and then pasting the picture into the text. The inserted picture can be sized and positioned as a single drawing element.

Inserting graphs from Excel

Two steps suffice to copy a graph from an Excel spreadsheet. First, display the extended Edit menu in Excel by holding Shift while clicking on Edit and use the "Copy, Picture" with the "As shown when Printed" option to copy the graph to the Windows clipboard. Second, use the "Paste Special, Picture" function in Word to paste the graph into the document. The "Float over text" option in Word should be deactivated for the graph to fit neatly between lines of text. This technique was used for most of the examples in this section with uniformly good results. The option of copying the graph as an Excel object—and not as a picture—is no easier and usually generates an image with slightly less resolution and general quality. An occasional problem occurs when the plot area in an Excel graph is needlessly defined to be transparent with no area fill color. This specification causes some printers, such as the Lexmark, to print a solid black graph area. Avoid this problem by merely leaving the plot area color on Automatic, which gives a white default area color. Apparently, it is handy to insert or paste a graph above at least one line of text. Otherwise it may be difficult to continue with text.

Inserting graphs from Quattro

A similar result can be obtained in Quattro by selecting the graph for editing or by selecting an embedded graph and copying to the clipboard. Then paste as a picture into the Word document. As with an Excel graph, the "Float over text" option in Word should be deactivated for the graph to fit neatly between lines of text. The graph can be pasted into the document by the same method used for an Excel chart.

Inserting separately printed graphs

On occasion, graphs will be generated by stand-alone instruments or data acquisition systems such as oscilloscopes or anemometers that do not produce computer files but only hard copies. The very best practice in such cases is probably to generate a high resolution image with a scanner and insert the image as a picture into the report. When this method is awkward or inconvenient, insert the separately printed graph as a separate page after the page in which the graph is mentioned. The best practice is to print the caption and page number on an otherwise blank report page and to very neatly paste the separately printed graph onto this page. Otherwise, it is marginally acceptable in practice, and fully acceptable for undergraduate lab reports, to paste a machine printed caption and page number onto the separately printed graph and insert this page into the report.

Comparative Examples

Compare the following examples, Figures 8.3 and 8.4, for clarity and effectiveness. Figure 8.4 has several basic deficiencies. The markers used to identify the literature model are misleading. They could be interpreted as data points which, very unlikely, exactly fit the model curve from the literature. The *x* axis is faulty. Since zero Celsius has no fundamental significance, there is no purpose in starting the horizontal scale there. The space left of 10°C is merely wasted. The extra title on the graph area in this figure is redundant. It is probably located correctly only for its intended use as a projection transparency. For a graph in a report, use a caption properly located below the figure. Either choice of vertical axis may be appropriate depending on the features of the data that are to be emphasized. Figure 8.2 emphasizes the difference between the data sets while Figure 8.3 tends to emphasize the reasonably good agreement.

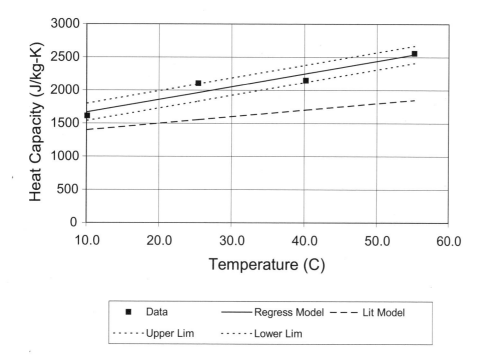

Figure 8.3. Heat Capacity Data (markers) and Regression Model (solid line) with Error Band (dotted lines) for Lubrication Oil Compared with Model from Tempco (1992, broken line)

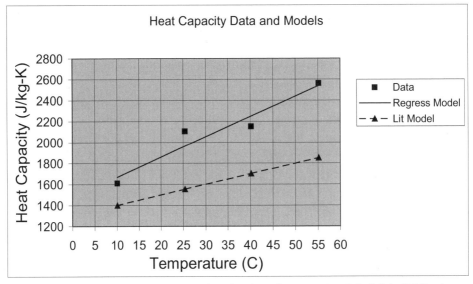

Figure 8.4. Heat Capacity Data (markers) and Regression Model (solid line) for Lubrication Oil Compared with Model from Tempco (1992, broken line)

The location of the legend block and a few other features in Figure 8.4 are faulty. Since horizontal space is limited, the legend seriously crowds the graph. If feasible move the legend to an unused area of the plot plane. Otherwise, move any legend block to the bottom. The gray tone printed on the graph area, which is the original default in Excel, seems highly undesirable. It makes the graph hard to read even in the original, much more so in a photocopy. It is always desirable to eliminate this tone. The grid lines in each graph may or may not be desirable. Certainly, the wider spaced gridlines in Figure 8.3 are less disruptive than the tight grid in Figure 8.4. Consider omitting gridlines entirely. Not incidentally, most of the formatting problems in Figure 8.4 can be completely avoided by using the recommended default graph described in the next section.

Setting the Default Graph

The developers of Excel have programmed the default graph to be a bar chart that is totally unsuitable for most engineering applications. To define a more suitable default chart, first obtain a prototype chart that suits your typical engineering application. A suitable prototype is available from the web page of one of the authors (Jeter 2002) or the user can set up an alternative with the instructions presented below.

An XY chart with lines and markers for all six series is usually most appropriate. Change to lines or markers or occasionally keep both when you use this graph type in actual applications. It is probably best to use all straight line segments, not a smoothed line, initially. All the lines and markers should be black in color and distinct in type. The types range from solid through the broken designs. The first series should probably have the simplest patterns, such as the unbroken plain line with square solid black markers. Set up the other features to suit your needs and preferences. There should be no title on the graph plane. The axis titles should be about 16 point in regular, not bold or italic, Arial font. The axis numbers and legend notes should be about 14 point. Remember these font sizes are as specified in the chart properties. The lettering will appear in the specified size only when the chart is printed alone. When the chart is pasted into a report the lettering will be smaller; therefore, relatively large fonts are specified in the properties. When finished, the prototype graph should look like the next figure.

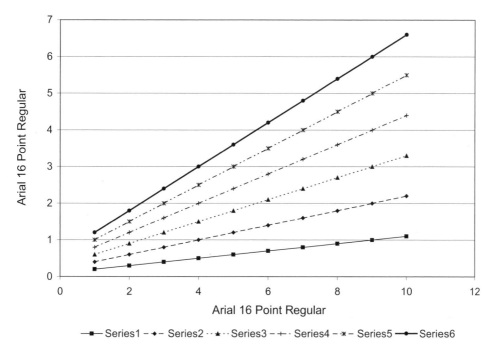

Figure 8.5. Representative Dummy Graph Created to be the Default Graph Type

The axis formats, both numerical and scaling, will always require some special attention. The Number format, which displays a fixed number of decimal places, is usually the best format in engineering practice. Unfortunately, without knowing the likely magnitude of the axis values you cannot set in advance the number of decimal places needed. For example, a value of .005 will display as 0.00 in a two place Number format. Consequently, you should specify the indefinite General format initially. After inspecting the results, always switch to a Number format with the appropriate and uniform number of decimal places. Similarly with no way to have advance knowledge of the possible range of values, the maximum axis value cannot be set in advance, so leave this value set to automatic. In contrast, it may be a good idea to set the minimum value to zero. This specification will always force you to give some consideration to all the implications before the automatic scaling is allowed to truncate an axis. These steps should complete an acceptable dummy graph.

Next carefully inspect the dummy graph, preferably after pasting it into a document. Make any further necessary or desired changes. Once you are satisfied with the dummy graph, you can specify it to be the prototype of a user-defined chart type. Then the new user-defined type can be specified to be the default type. This overall operation takes several steps. First define the new type as follows. To begin, ensure that the chart is the selected item in the worksheet, and then pull down the Chart menu on the main menu bar. Then select the Chart Type option on the Chart menu. Next select the Custom Types tab. Select the User Defined option. Punch the Add button and complete the Add Custom Chart Type dialogue box giving this chart type a descriptive name such as "typ-xy." The dialog box should now look like the Figure 8.6. When finished adding this new user-defined chart type, make it the default by punching the Set As Default Chart button. The newly created chart type will now be the default.

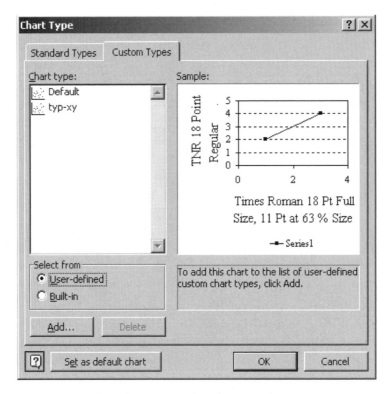

Figure 8.6. Dialog Box for Chart Type After the Typical XY Type Has Been Defined

It takes a few minutes at most to set up your preferred style of graph as the default, but you will save much time later. A spreadsheet called Default should be available on the web page maintained by one of the authors (Jeter 2002). This spreadsheet includes a graph of the type illustrated in Figure 8.5. You can use this spreadsheet to set the defaults on your home computer. Unfortunately, convenient defaults can not be reliably set up and maintained on any computers in a public cluster, so you may want to keep a copy of Default.xls on the diskette or other portable medium where you keep your data and reports.

You can create a chart of the default type on a separate sheet in only two steps by selecting the data and hitting the F11 key. When the graph appears on a separate sheet, all of the patterns will be precisely as you have specified. Alternatively, you can create the chart as an embedded object in any sheet by selecting the data and running the chart wizard. Unfortunately, when the graph is first inserted as an object on a worksheet some of the features, especially the font sizes, will be distorted. You must correct these distortions to create a high quality embedded graph. These corrections can be tedious. Indeed, working with the chart wizard is so awkward that it is surely easier to just create the graph on a separate sheet and then copy and paste it to a worksheet. Either way a good quality graph can be created almost instantaneously by specifying convenient defaults.

Summary Table

Table 8.1 summarizes the guidelines in this section and can be used as a guide to grading graphs in undergraduate lab reports.

Closure

Graphs are obviously among the most important components of any technical report especially an experimental engineering report. Please strive for a professional result with clear content and excellent production qualities in every graph.

References

Jeter, S. M., 2002, "Example Spreadsheet, Default.xls," ME 4053 Engineering Systems Laboratory, the George W. Woodruff School of Mechanical Engineering, Georgia Institute of Technology, Atlanta, GA, 4 January; document is available online at <me.gatech.edu/sheldon.jeter>.

Table 8.1. Guide to Preparing Graphs for Experimental Engineering Reports

Design is neat, uncluttered, legible, and effective.	Unique number is assigned and descriptive caption is given.
Always place caption below graph in written report.	Consistent capitalization system used in all captions.
XY graph, not Line chart, is almost always used.	Graph box included only with discretion and then consistently used in one report.
No reliance on color or subtle gray scales.	No shading of graph area.
Graph is cited in text.	Any cited graph is included.
Graph is inserted into text at first feasible location when possible or on separate page when necessary.	Graph is attached (*i.e.*, as a one item appendix) to extended abstract only.
Full page graph on separate page in landscape orientation has caption to right.	Separately printed graph is pasted onto page or has machine printed caption pasted on to graph page.
Zero point is purposefully and appropriately included or excluded in axes.	Secondary axis is included when necessary.
Axes are not so compressed or span is not so great that variation in data is not apparent.	Special care taken not to obscure scaling relations between normalized variables.
Axes are titled, with dimensions if existing.	Markers are strictly reserved for discrete data.
Data is presented in scatter plot or connected profile of markers as appropriate.	Distinct line types (*i.e.*, not markers) are used to identify continuous data or model curves.
The series in a single series graph is identified only in caption.	All series are identified in legend, in notes on graph, or in caption.

Chapter 2.9

Illustrations

Engineering reports frequently include illustrations such as drawings and pictures. A complex drawing or especially a photograph may be difficult to interpret. Complex drawings include engineering assembly drawings and artistic line drawings.

Simplicity is usually the most desirable feature in any figure. Any simple drawing, especially a schematic, such as Figure 9.1, will be better understood than a photograph, such as Figure 9.2, or a more complex drawing.

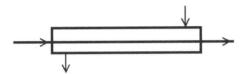

Figure 9.1. Schematic of Crossflow Exchanger Element

Even the rather complex schematic in Figure 9.3 is much more comprehensible than the photograph of the apparatus in Figure 9.2. A schematic may be better than an engineering assembly drawing or even an artistic drawing. Artistic line drawings can be very effective but require skill, time, and expense. Usually, avoid using a so-called engineering "detail" drawing, such as Figure 9.4, in a report. A detail is the name for a shop drawing with dimensions and manufacturing notes. Construction or manufacturing drawings are usually presented separately as an attachment if a single item or in an appendix or separate bundle. If necessary, simplify a detail drawing if it must be used. Figure 9.5 is an example. As in the example, deactivate all layers of an AutoCAD or similar drawing except the layer containing the visible or so-called "body" lines. Be especially critical of any proposed pictures. Neither a scanned photograph nor the typical digital image has high resolution, and the usual color photo or image loses much informative value when transformed to gray scale. Usually a simplified drawing should accompany or even replace a photograph of ordinary quality. See the summary table following the example figures for recommended guidelines.

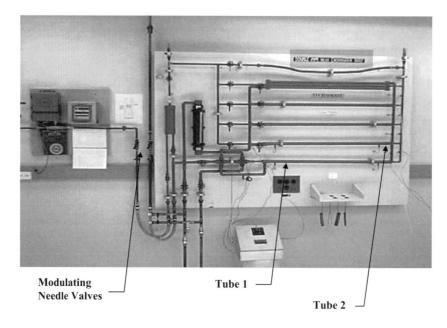

Figure 9.2. Photograph of Heat Exchanger Apparatus

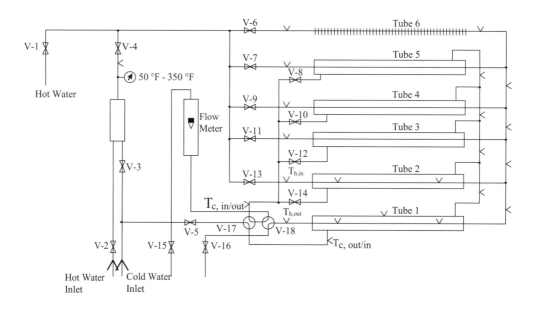

Figure 9.3. Schematic of Heat Exchanger Apparatus

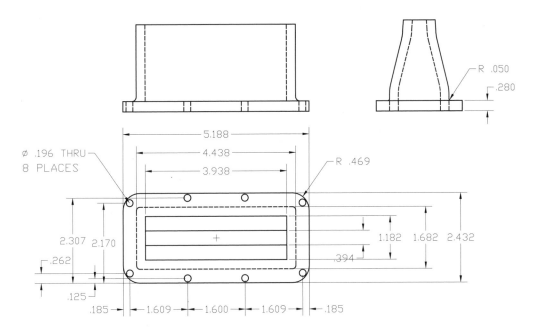

Figure 9.4. Engineering Detail Drawing of Experimental Nozzle

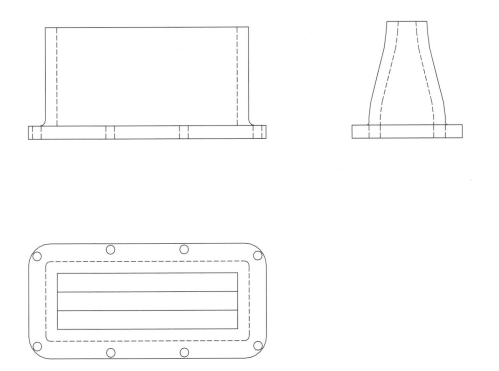

Figure 9.5. Simplified Drawing of Experimental Nozzle

Summary Table

Table 9.1 summarizes the guidelines in this section and may be used as a guide to presenting illustrations in undergraduate lab reports.

Table 9.1. Guide to Preparing Illustrations or Figures other than Graphs
for Experimental Engineering Reports

Figure is neat, uncluttered, legible, and effective.	Unique number is assigned and descriptive caption is given.
Caption always placed below figure or illustration in written report.	Consistent capitalization system used in all captions.
Photograph used only with discretion and usually accompanied by schematic.	Usually a schematic or simplified drawing is preferred.

Chapter **2.10**

Guidelines for Equations

Introduction

Mathematical equations were until recently the bane of writers, typists, and printers, but modern equation editors have eliminated most of the difficulty. Ensure that the equation editor is installed on your personal system. If the equation editor has been installed, then it will appear as an option in the Insert Object dialogue box in MS Word. Otherwise it will be necessary to install this missing component. The suggestions in Chapter 2.15 on computer usage should work for users of the Microsoft Office suite of programs. The Excel or Office program disk will be needed. The procedure for adding the editor to a separate installation of the MS Word program will be similar but different in minor details.

Grammatical Role

Grammatically, an author has two alternative and acceptable ways of treating an equation: consider it as a single mathematical entity with the role of noun or noun phrase; or consider it as a statement where the equality or inequality sign serves as the verb. Write the accompanying text so that the equation corresponds to its grammatical role, in the first case as a phrase in a sentence or in the second case as a clause in a sentence or as a separate sentence in a paragraph.

Punctuation, Numbering, and Placement

Grammar and logic aside, the current and unfortunate style is to omit any terminal punctuation that would seem needed to follow an equation, such as

$$F = kx = m\ a, k = cons\tan t \qquad (10.1)$$

where, $\quad F\ =\ $ force $\qquad\qquad k\ =\ $ spring constant
$\qquad\quad x\ =\ $ deflection $\qquad\quad m\ =\ $ acceleration
$\qquad\quad A\ =\ $ acceleration

Note that the equation is easily centered using the ⊥ tab and that the equation number is right justified with the ⌐ tab. To allow an equation to be positioned properly, **always disable** the "Float over text" option when inserting MS Equation 3.0 in MS Word. Only a few other simple rules for integrating equations into text are really necessary. For instance, usually an equation should not end a paragraph, and a series of equations should not be presented without intervening textual comments and explanation. The details of equation editing are similarly simple, but a few guidelines that can save time and effort and minimize frustration are presented in the next section.

Entering and Editing

Equations are easy with a sophisticated equation editor such as the one included with MS Word. Typically an equation editor creates a frame or a short embedded document. The frame contains the information to generate the equation, probably in the form of a typesetting language such as TEX. While highly developed, this and similar editors do have a few confusing or awkward characteristics. These bugs can be problems to a novice, but as described in the next several paragraphs, they are easily managed once understood.

Math and text modes

By nearly universal convention, math symbols in an equation, like text and tables, should be in Italic type. Consequently, Equation Editor assumes that most symbols are mathematical unless it recognizes a function name. The raw result of the defaults is shown in the poor example on the previous page. Instructions are given below on how to convert notes, such as "sp" and "constant," to text style as in this improved example,

$$F_{sp} = kx = m\,a, \; k = \text{constant} \qquad (10.1)$$

For consistency and clarity, use italic type in the paragraph text when referring to a math symbol such as F_{sp} for the spring force.

Overriding text or math defaults

Equation Editor operates in either a "text" mode or a default "math" mode. As mentioned above, the text mode is infrequently used such as to add a notation directly to an equation. Text mode is much like a conventional text editor with standard characters and spaces inserted by the spacebar. In math mode, equation editors are essentially typesetting programs and usually space the characters correctly by default. When necessary, extra spacing can be inserted manually as was done in the second equality in Equation 10.1. To insert an extra thin space, use the Ctrl-space combination. In math mode, most characters are assumed to be variables and are in italic type. When the abbreviation for a common math function, such as the tangent, is encountered, the editor switches from italic to standard type (*e.g.*, "*ta*" will be finished as "tan") in accord with standard practice. This feature causes the word "*constant*" to appear in the first version of Equation 10.1. To switch to standard type, select the material such as a word or characters and convert it from math to text. Use the pull-down Style menu as illustrated in Figure 10.1 below for this conversion. The style menu will be an item on the main menu at the top of your screen when Equation Editor is activated. For example, the notation in Equation 10.1 appears as "*constant*" in math style. To fix this problem, select the text with the cursor and give it the Text property by pulling down the Style menu and selecting the Text mode. The notation will appear properly as "constant" in text style. The same procedure was used to change the subscript "*sp*" to "sp" since the subscript is not a variable name but merely a notation standing for "spring."

Parentheses

It is definitely preferable to select the parentheses template from the Equation menu rather than to type the parentheses characters from the keyboard since the template will always automatically resize and reshape itself to fit its contents while the characters are fixed in size and shape. Otherwise you can have awkward results such as $(a - \dfrac{1}{2})$ rather than $\left(a - \dfrac{1}{2} \right)$.

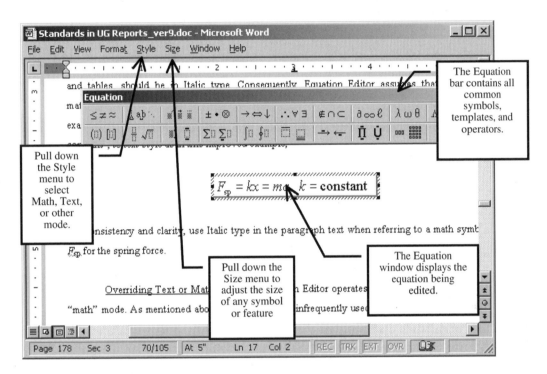

Figure 10.1. Document Window Showing Open Equation Window, Equation Menu, and Location of the Style and Size Pull-Down Menu Items on Main Menu

Inserting extra spaces

In math mode, Equation Editor is essentially a typesetting program and usually, but not always, spaces the characters correctly automatically. To modify the default spacing, Equation Editor has several complicated features for adding spaces between and fixing the alignment of equation elements. While ingenious, most of these complex features will never be needed. If necessary, these features can be accessed from the obscure-looking menu item in the second position of the first line of the Equation menu. See Equation Editor Help for information on these items. In almost all practical cases, extra spacing can be inserted more easily by simply typing the Ctrl-space combination to insert an extra thin space. This method was used to modify the second equality in Equation 10.1. Text mode is much like a conventional text editor with standard characters and spaces inserted by the spacebar.

Narrow equation window

On occasion, the equation window does not open far enough to fully display the entire expression, and the initial or final character will not be printed even though it appears on the monitor. Presumably, this feature is a bug and will be corrected in future versions of Word or Equation Editor. However, an easy way to avoid this problem is to always begin and end every equation with a thin space inserted with the Ctrl-space key combination. This extra step requires little effort when an equation is being written and eliminates a problem that will only be evident after the document is printed.

Stuck character property

Sometimes when the equation ends with a subscript or superscript, this character property remains in force even after the equation editor is closed. This feature, which is hard to reproduce, can cause the next text in the document to mysteriously appear out of usual alignment. To correct this problem, insert a thin space at the end of the equation on the main line of the equation. This space character will unobtrusively deactivate the super- or subscript function.

Sizing or resizing

Finally, some users experience difficulty resizing an equation such as changing from 12 points in a document to large type for a presentation slide. An equation window looks like a drawing object, but it really is a frame. A drawing object can be resized by first selecting it and then pulling on one of the markers called a "handle," which appears when the object is selected. In contrast, an equation does not always expand reliably when the size of its box is increased. Even when the equation does expand, maintaining a consistent type size from equation to equation may be difficult. A better approach is to reset the size parameters for the equation editor. The size menu is accessed on the menu bar, at the top of the Word screen, when the equation editor is active as shown in the figure above. For example, Equation 10.2 was written with the symbol size defined as 24 and the superscript defined as 14. This font is about the smallest size acceptable on a presentation slide.

$$e = mc^2 \tag{10.2}$$

Note that if Equation 10.1 were now edited with the size parameters reset to the default values, it would be resized automatically. Use the "default" option on the size menu to return to the standard sizes. When only one or a few characters need to be resized, select the character and use the "other" option in the size menu as was done to resize the "e" symbol to 36 points in the next equation.

$$\textit{\Large e} \; = mc^2 \tag{10.3}$$

Beware when mixing size definitions in one document. If the default size is changed, every equation that is subsequently opened will be resized.

Fine adjustment or alignment

The more recent editions of Equation Editor have some very versatile and useful fine adjustment and alignment features. For example, the "nudge" command, which is implemented with control key plus the appropriate cursor key, was used to adjust the position of the exponent a bit higher and lower than the central default position in the following equation.

$$e \; = \; mc^2 \; = \; mc^2 \; = \; mc2 \tag{10.4}$$

The alignment symbol is a non-printing symbol that resembles a delta, Δ, in the ellipsis template. It is very helpful in maintaining precise vertical alignment in multiple line equations. If you need the flexibility of these commands, refer to the Equation Editor help menu.

Dreaded "Disk full error"

One persistent fault of the MathType equation editor is an apparent communications lapse with Word that creates this annoying and somewhat disruptive error. Even the text of the message is misleading. The disk is not really full. Apparently, Equation Editor fails to properly close the embedded object. Word interprets this failure as creating a document of indefinite length that cannot be closed. Windows then misinterprets this indefinite length as infinite length. In technical terms, Word and Equation Editor start a fight, and Windows, instead of suppressing it, urges them on. The disk may not be full, but the document is disrupted.

This error cannot be avoided, but major repercussions can be. It is good practice to save often when editing many equations, especially long ones. Save before starting the equation editor and after finishing. Then if the error occurs, it should be limited to a problem with the last equation. Fixing the problem can be straightforward if it is limited to the last equation. Try to open the last equation. If it can be opened, close it and try to save the file. If the file can be saved, the problem is fixed. If not, open the equation and try a small modification such as adding a space at the end. Then try to save the file. Again, if the file can be saved, the problem is fixed.

If the last and presumably faulty equation cannot be fixed, more drastic action is required. Assuming the problem is limited to the last equation, delete it. Then it should be possible to save the file. After saving the file, the deleted equation must then be freshly entered.

If the file cannot be saved after the last equation was deleted, the problem is more complex. Assuming that only one bad equation exists, work back through the document. Delete each equation individually in turn, and try to save the file. If the particular equation just deleted was not the problem, it will not be possible to save the file, so restore it with the back-edit arrow. When the one faulty equation is deleted, it should be possible to save the file. With patience, this technique should work unless two bad equations were created.

In the very complicated extreme cases, the error will cause the loss of all information since the last successful save. Luckily this error is not too frequent, but do save often. Do not despair if the error does occur, since the method described above can save most of your work.

Obviously, equations are characteristic and important in technical writing; therefore, construct them carefully and give them appropriate development and attention in the text.

Summary Table

Table 10.1 summarizes the guidelines in this section and will be used as a guide to grading equations in undergraduate lab reports.

Table 10.1. Guide to Writing Equations in Experimental Engineering Reports

Equation has logical grammatical setting.	Equation is neat, uncluttered, legible, and correct.
Equation is generated with Equation Editor not a line of text.	Equation is centered on separate line with unique number at right margin.
Equation is proper size, especially for projection slide.	Equation box is sufficiently wide to avoid truncation.
Templates from menu are used for parentheses, not symbols typed from keyboard.	Math symbols are in italics with functions, numbers, and notes in regular type.

CHAPTER 2.11

GUIDELINES FOR SPREADSHEETS

Introduction

Systematic organization of a spreadsheet or multiple page workbook will help make your numerical and graphical work easier and more reliable, and your spreadsheet will be easier to read or review or modify. Entire spreadsheets or even large blocks are usually too expansive and informal to be included in formal professional reports. For formal reports, you should extract a concise block of the spreadsheet and incorporate the block into your report as a table.

Even though a spreadsheet may not make a suitable exhibit, it must be organized well enough to produce reliable results. Organization is especially important if the spreadsheet must be reused or shared or used by others. In addition, spreadsheets are commonly included in less formal reports and most undergraduate lab reports to facilitate the monitoring of the data and calculations. Be sure to cite this exhibit in the text. It is usually called an attachment, which is a one item appendix. An entire brief spreadsheet, or a pertinent block, may also be printed out and then physically pasted into a research notebook to create a secure documentary record.

To create an effective spreadsheet, organize the presentation functionally and design and format the entire spreadsheet or at least its functional blocks so that it can be edited and updated efficiently and printed or inserted neatly. The organization described below and illustrated in the accompanying example has been found to be useful for most experimental work.

Functional Blocks

Organization into the four sections described below and illustrated in the accompanying example has been found to be useful and effective for most experimental work. This organization is required in undergraduate lab assignments.

Header area

This very short section should include the name of the experiment, the names of the experimentalists, the course number and section, the time and date on which the experiment was performed, and the file name. Be sure to give enough information to identify the purpose of this file.

Summary section

This typically short section should include only the most important summary results such as numerical value for an overall flow or a critical performance variable. Put this block near the top of the spreadsheet, so it will not be lost in a large spreadsheet or workbook.

Preliminary section

This section of medium length should include a list of constants, geometric information, and conversion factors that are used in the calculation of intermediate and/or final results along with any non-recurring calculations used in the data processing or presentation. **Importantly**, put all unique data such as experimental parameters, atmospheric pressure, calibration constants, and so forth here so that this data can be comprehensively updated if necessary.

Data and calculations

This section should include the experimental data and the repetitive calculations for data processing or presentation. Put only routine data here. Use cell references to unique data placed in the Preliminary Section. Don't hide unique data here that will make the spreadsheet hard to correct or update. This section may include an embedded graph of the data that one may or may not choose to include in any attachment appended to a report. It is probably good practice to embed a graph of the data when a block of the spreadsheet is printed to be physically pasted into a research notebook.

Placement of sections

Obviously, the header should be at the top of the spreadsheet. The summary and preliminary section should probably come next, but the order of these two sections is immaterial. The data and calculations section should probably come last to make it easy to extend this block if necessary.

Detailed Guidelines

Inserting labels

Rows and columns should be labeled appropriately to identify data or results. In addition you should list the units in parentheses below or beside the column/row headings. Do not mix unit labels with your data. It is better to restrict the units to the column and row labels.

Identifying data

A handy technique for identifying data is to place the numerical value in the left most column in a block and insert the identification in the column to the right. The identifier should be left justified. The identifier entry must begin with a single quote to identify the entry as left justified text not a cell formula. This technique makes a neat and uniform alignment easy, and the identifiers will be left justified like ordinary text as shown in Table 11.1.

Obviously this method is efficient, and it may look neater than placing the identifiers to the left and making them right justified. Note that the identifying phrase has a leading single quote to identify it as text not a formula. This technique is used in the example spreadsheet attached to this section.

Table 11.1. Example of Aligning Identifying Phrases in a Spreadsheet

Data as Entered	Identifier as Entered
1.05 (right justified)	' = scale
0.02 (right justified)	' = offset (kPa)
Data as Displayed or Printed	Identifier as Displayed or Printed
1.05	= scale
0.02	= offset (kPa)

Formatting numerical data

To facilitate editing and to produce a neat appearance, use a consistent format for every column of data or block of related data. As shown in Attachment 11.1, all of the temperature data in column A are printed in Number format with one decimal place. Similarly, all of the vapor pressure data in columns B through F are printed in Number format with no decimal places showing. As is typical with Windows applications, access the format by selecting a block and right clicking. In Excel as in Word, the format, including the numeric format, can be transferred from a formatted block with the format painter.

Displaying decimal places

In a table, it is important to avoid displaying trailing insignificant digits. A spreadsheet is less formal, and it may be desirable to display trailing digits to track calculations or maintain consistent formats. Even so, it is good practice not to display an egregiously long tail of insignificant trailing digits. It is practically impossible to display only significant digits in a spreadsheet. The numbers cannot be truncated without changing the results of calculations and introducing possible numerical error when calculations are repeated. The Round worksheet formula can be used, but it is tedious and awkward.

Formatting for printing

Use good judgment in formatting and printing a spreadsheet or a portion of a spreadsheet. Only when required should you print the entire contents of a spreadsheet, and then you should append it to your report as a single item appendix and call it an attachment. Such extensive attachments would not be acceptable in formal professional practice except for internal use. A similar use is in an undergraduate lab course displaying data and intermediate calculations. Typically, one should extract a small and specifically designed block of only pertinent data from the spreadsheet and refine and present the block as a table. If a spreadsheet attachment is used, it should fit onto a single page or on a single page with continuing pages that have their own adequate organization. Do not print continuing pages of incomprehensible hanging columns or rows. To conserve space, you may use 10 point type or the "Print to Fit" option, but never print with a type size too small to be legible. Even in a working spreadsheet, it is desirable to avoid displaying excessive insignificant digits; however, it may be desirable to print a few extra digits to help one trace intermediate calculations. Even so, never *report* excessive insignificant digits in a table or even a spreadsheet block intended for general circulation. In such cases, use reasonable data formats and the Round function or manual rounding to generate blocks that are consistent with the uncertainty of your measurements.

Printing spreadsheet as separate item

The spreadsheet, or at least its important blocks, may be printed as a separate item and appended to the report as a separate attachment. Include the unique attachment number and descriptive title in a header printed above the spreadsheet block.

Inserting spreadsheet as a picture

A convenient way to generate an attachment that contains a large spreadsheet block is to copy the block as a picture and add the picture to the report as an attachment. Three steps suffice to copy an Excel spreadsheet as a picture. First select the spreadsheet, or at least its important block, for the picture with the cursor. Next use the copy picture option. Display the extended Edit menu in Excel by holding Shift while clicking on Edit and use the "Copy, Picture" with the "As shown when Printed" option to copy the spreadsheet blocks to the Windows clipboard. Then use the "Paste Special, Picture" function in Word to paste the picture into the document. The "Float over text" option in Word should be deactivated for the picture to fit neatly under its header. When the "Float over text" option for an object is deactivated, the object appears as an "in-line" frame rather than a "float over text" frame. An in-line frame acts like a single text character. It holds its place in the text and responds to the alignment features. This technique was used for most of the examples in this section with uniformly good results as in Attachment 1 below. As with graphs, it is apparently good practice to insert or paste a picture above at least one line of text. Otherwise, it may be difficult to continue with text. The picture of the spreadsheet should be preceded by a heading in the document with its unique attachment number and descriptive title.

Printing cell formulas

Only when required, submit a set of representative cell formulas with your lab report. This listing must be the formulas for a concise block representing a set of crucial calculations such as a numerical integration. To display cell formulas for printing in Excel, click on the Tools item in the standard toolbar. Then select Options. In the Options box, select the View tab and in the Windows Options area select the Formulas option. Always include such a block showing formulas as a separate attachment such as Attachment 11.2. Identify this item with a unique attachment number and a descriptive title, and always describe the block of calculations in the accompanying text.

Regression block

The regression block generated by Excel is extensive and awkward to format. The regression block should probably be placed on a dedicated sheet of the workbook. Additionally as shown in the example, it may be desirable to extract a concise block of the regression and place that block in the main part of the workbook. Use cell formulas to relate the concise block to the original regression output. Relegate the Excel regression outputs to separate pages of the workbook.

Examples

The simple examples, Attachments 11.1 and 11.2, show the unique identifying number and descriptive title, printed as a header. Also shown are the heading, the preliminary section including general data, the summary section, and the section of routine data and calculations in the spreadsheet itself. Note that the title is printed as a header so that it will appear centered and prominent. Use of the header is especially desirable to display the heading on every page when a multiple page printout is needed.

Summary

Organize spreadsheets into four functional and logical blocks for the heading, the summary, the block of unique data and one-time calculations, and the block of routine data and repetitive calculations. A working spreadsheet is usually too extensive and too informal to be included in a widely published report. Instead, extract pertinent data into concise tables and integrate the tables into the text. In a less formal report, include a spreadsheet as an attachment that may be printed separately or preferably inserted as a picture attached to the main report. When a multiple page spreadsheet must be attached to a report, pay special attention to its organization for printing: print only the necessary blocks and avoid pages with dangling or poorly identified columns or rows. Pictures of spreadsheets or separately printed versions attached to informal reports, such as student reports, should include all important blocks and have a unique attachment number and descriptive title in the printed header. In all cases, the spreadsheet should be functionally organized, neatly formatted, and well identified. Table 11.2 summarizes the guidelines for spreadsheets as discussed in this chapter.

Table 11.2. Guide to Preparing Spreadsheets and Incorporating Them
into Experimental Engineering Reports

Overall design and detailed format are neat, uncluttered, legible, and effective.	Unique number is assigned and descriptive title is given in heading.
Spreadsheet is presented as an attachment cited in text.	Concise block of pertinent data is also extracted and presented in a table.
Title heading is centered and prominent with type size at least 10 point.	Spreadsheet is structured with heading, preliminaries, summary, and data blocks.
Heading identifies purpose and history of spreadsheet.	Unique data, constants, and parameters are placed in the preliminary section.
Summary section is near top of spreadsheet.	Routine recurring data are in a well-organized separate section.
Consistent numerical format is used in a related block or blocks of data.	Units are identified for every dimensional data block, typically in column heading.
Recurring data are identified, usually with column header.	Unique data are identified, usually with identifying phrase.
Printed margins are adequate and at least 1 inch all around.	Multiple page printout is carefully designed especially in continuing pages.
Block of formulas presented in separate attachment if required.	Optional concise summary regression block prepared if appropriate.

Attachment 11.1 Experimental Spreadsheet for Vapor Pressure Lab[1]

file: PVAPOR99   21 June 1999   S. M. Jeter
        Spreadsheet for processing vapor pressure data for CFC-12.

Summary Results
        0.9943424 = R squared
        0.8501812 = alpha risk

Constants and Parameters
        Calibration data: 1.0 = scale
                                         0.0 = offset (kPa)
                97.9 = atmospheric press (kPa)

Experimental Data

Temp C	Pv Data kPa	Pv Model kPa	Pv Lit kPa	Pmes kPa	Pcorr kPa	1/T 1/K	1/T^2 1/K^2	ln (Pv) Data
30.1	748	756	746	650	650	0.003298	1.087E-05	6.617269
45.2	1098	1061	1062	1000	1000	0.003141	9.867E-06	7.001155
59.9	1398	1452	1519	1300	1300	0.003003	9.015E-06	7.242726
75.0	1998	1971	2082	1900	1900	0.002872	8.250E-06	7.599852

Concise Quadratic Regression Block

Constant	17.094807	
Std Err of Y Est	0.053826	
R Squared	0.994342	
No. of Observations	4	
Degrees of Freedom	1	
X Coefficient(s)	-4229.962	320157.3
Std Err of Coef.	8243.672	1335225.3
t-stat		0.239778
alpha		0.850181

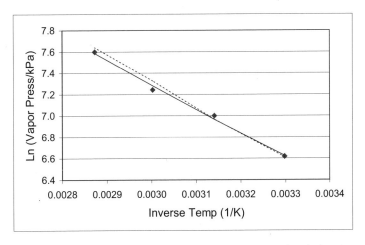

[1]Print the title as a header in the spreadsheet so that it will appear centered and prominent on every page, or insert the title as heading in the document above the picture of the spreadsheet in this example.

Attachment 11.2 Block of Experimental Spreadsheet Showing Formulas Used for Vapor Pressure Lab

file: PVAPOR99   21 June 1999   S. M. Jeter

Spreadsheet for processing vapor pressure data for CFC-12.

Summary Results

=Sheet4!B5   = R squared
=Sheet4!E19   = alpha risk

Constants and Parameters

Calibration data: 1   = scale
                                     0   = offset (kPa)
97.9   = atmospheric press (kPa)

Experimental Data

Temp C	Pv Data kPa	Pv Model kPa	Pv Lit kPa	Pmes kPa	Pcorr kPa
30.1	=F20+$B$13	=EXP(J20)	745.8	650	=$D$11*E20+$D$12
45.2	=F21+$B$13	=EXP(J21)	1062	1000	=$D$11*E21+$D$12
59.9	=F22+$B$13	=EXP(J22)	1519	1300	=$D$11*E22+$D$12
75	=F23+$B$13	=EXP(J23)	2082	1900	=$D$11*E23+$D$12

TABLES

Introduction

Tables allow the presentation of extensive quantitative information in a limited space. To be effective, tables should be carefully designed, well integrated, and concisely formulated. Table 12.1, recommended in Chapter 2.16 of this text for error propagation analysis, is a reasonable example. Note that by long-standing convention, the descriptive title is always above a table as shown in the accompanying example. The left hand column is usually the independent variable or a list of categories. The first row is usually an array of text entries called "heads" that identify the columns and frequently include units. Use nouns or noun phrases as heads. Pay attention to formatting and alignment. Frequently the heads are centered, and the quantitative data are either aligned by their decimal points or, if whole numbers, justified to the right. Be sure the data are internally consistent. A critical reader can easily check tabulated data, so be careful to verify the data and calculations. Never tabulate any trailing insignificant digits. Avoid a long narrow table; repeat the columns instead.

Table 12.1. Estimate of Bias or Uncertainty B in Shaft Power Measurement

Measurement	U_x [a]	Influence Coefficient, $\dfrac{\partial \dot{W}}{\partial x_i}$	$U_i^2 = \left(U_{xi} \dfrac{\partial h}{\partial x_i} \right)^2$	Basis	Source
shaft speed, N	10.0 RPM	$\dfrac{\dot{W}}{N} = \dfrac{7050\ \mathrm{W}}{1020.\ \mathrm{RPM}} = 6.91\ \dfrac{\mathrm{W}}{\mathrm{RPM}}$	4800 W^2	resolution	(1)
arm length, r	2.0 mm	$\dfrac{\dot{W}}{r} = \dfrac{7050\ \mathrm{W}}{307\ \mathrm{mm}} = 23.0\ \dfrac{\mathrm{W}}{\mathrm{mm}}$	2100 W^2	measurement	(2)
force, F	5.0 N	$\dfrac{\dot{W}}{F} = \dfrac{7050\ \mathrm{W}}{215\ \mathrm{N}} = 32.8\ \dfrac{\mathrm{W}}{\mathrm{N}}$	27000 W^2	calibration	(3)
		sum of $U_i^2 =$	33,900 W^2		
		Expanded Uncertainty B [b] $=$	180 W		

Sources: (1) physical inspection, (2) precise measurement, see text, (3) calibrated by manufacturer (1997)
[a] Expanded Uncertainty B in individual direct measurement, [b] 95% confidence limit

Design and Editing
Numerical formats

The numerical data must be in a format consistent with the number of significant digits. Usually a consistent format (*i.e.*, fixed number of decimal places in fixed point or scientific notation) can be used within a column, but you should break the pattern if necessary to avoid displaying insignificant digits. Never display a tail of insignificant digits. Use scientific notation for very large or very small magnitudes.

Numerical format for statistics

As discussed in Chapter 2.16, statistics are indirect measurements or numbers calculated from more direct measurements. It is possible to evaluate the uncertainty in any indirect measurement from the uncertainty of the underlying more direct measurements, and these uncertainties should always be known. However, this evaluation can be complex and is usually avoided except for the parameters in regression models. Therefore, the actual number of significant digits in other familiar statistics such as the R-Squared or the alpha risk in regression analysis is usually not known. Numerical analysis shows that these statistics can have many significant digits even when the data have few significant digits. As an expedient, it seems reasonable to present statistics with indefinite uncertainties such as the R-Squared or the alpha risk with enough digits to make any necessary comparisons. For example, use enough decimal places to distinguish two R-Squared values (*e.g.*, .9996 < .9998) even if the underlying direct measurement data have fewer significant digits.

Gridlines

Most tables should have gridlines to separate the columns and the rows. In Microsoft Word select the entire table and use the preset called "Grid" in the Format, Borders and Shading pull-down menu to draw these or more complex gridlines.

Table footnotes

As illustrated in Table 12.1, tables may be immediately followed by footnotes. In particular, an overall source footnote to acknowledge the origin should follow any table that is entirely or mostly borrowed. Similarly, the source of data for any part of the table can be attributed with a more specific source footnote. Engineering tables frequently include very brief explanatory notes actually within the table. Longer explanations can be written in a footnote or be included in the text with a reference in a footnote. Such source or explanatory footnotes that refer to parts of the table may be indexed with a capital letter in parentheses as illustrated in the example above. Footnotes explaining or identifying a specific entry may be indexed with a lower case letter in superscript as also shown in the example. Letters are preferred to numbers as indices to table footnotes, probably to avoid confusion with exponents, page footnotes, or entries in the list of bibliographical references.

Cell format

Typically a table begins as a block in a spreadsheet that is then copied and pasted into the report document. As shown in the first column in example Table 12.2 below, the resulting format has the numbers flush to the right. This alignment is only marginally acceptable. A somewhat better alignment is used in the second column where the data is centered in the text using the toolbar icon. An even better alignment is used in the third column. Here, the Format, Paragraph box has been used to define a right indention of 0.1 inch. This format leaves a little space to the right of the number. Either alignment in the second or third columns is acceptable but not optimal. Note the results in the fourth and fifth columns. In both columns, the number of decimal places in the last three rows has been adjusted to avoid displaying insignificant digits, so now the decimal points no longer are in alignment. The last column is optimal. Here, the decimal tab is used to align the decimal points. This standard tab symbol looks like an inverted tee with an accompanying decimal point or ⊥• approximately. This tab can be inserted into the ruler by the usual procedure. Start with the numbers in the cell aligned left. Then the decimal points in the numbers will automatically align on the tab stop. Other tabs in table cells are rare. If a tab stop such as the usual left tab is needed in a table cell, insert it in the ruler. To insert the actual tab in a cell use the Ctrl-Tab key combination as the tab key alone merely shifts the cursor to the next cell.

Table 12.2. Examples of Various Formats in Table Cells

angle(rad)	angle(rad)	angle(rad)	angle(rad)	angle(rad)	angle(rad)
3.1416	3.1416	3.1416	3.1416	3.1416	3.1416
6.2832	6.2832	6.2832	6.2832	6.2832	6.2832
12.5664	12.5664	12.5660	12.566	12.566	12.566
25.1328	25.1328	25.1330	25.133	25.133	25.133
50.2656	50.2656	50.2660	50.266	50.266	50.266
default from spreadsheet	centered	right indent paragraph format	centered	right indent paragraph format	decimal tab stop used

Use of the Table utility

The Table utility in MS Word is very useful and has only a few awkward or obscure features. It is a bit awkward to center a table on a page using older versions of Word. One must place the insertion pointer within the table and with the mouse cursor select (1) the entire table using the Table pull-down menu, (2) the Cell Height and Width menu item, and (3) the Row tab. Then center every row with the Center option in the Alignment block. In Word97 and later versions, one may alternatively select the entire table and center it with the button on the formatting toolbar. Also note that the ordinary tab key merely shifts the insertion pointer to the next cell. To insert a tab within a cell, use the Ctrl-tab combination. Neat and informative tables, such as the example, are easy to create in MS Word and definitely enhance technical reports.

Summary Table

Table 12.3 summarizes the guidelines in this section and can be used as a guide to editing and grading tables in undergraduate lab reports.

Table 12.3. Guide to Generating Tables for Experimental Engineering Reports

Table is cited in text.	Table is concise, neat, uncluttered, legible, and internally consistent.
Table has unique number and descriptive title in heading.	Table is centered horizontally on the page.
No insignificant digits are displayed.	Data is identified usually in column headers with units as existing.
Table is limited to one page.	Numbers are properly aligned.
Consistent numerical display format used except as limited by significant digits.	Scientific notation used for very large or very small magnitudes.
Appropriate text format, usually centered, is used for headers and notes.	Source and descriptive notes and footnotes are provided.

CHAPTER 2.13

LISTS

Technical writing can often be better organized and condensed by using lists. Two types of lists are common: "vertical" lists in column format, and lists in text style called "run-in" lists. Vertical lists are common in this manual. Both types help organize and simplify technical writing.

A run-in list is merely a series within a sentence or paragraph with the items numbered in parentheses and separated by commas or periods with a final conjunction (*e.g.*, "and" or "or") as in ordinary writing. Insert minimal or no additional punctuation in such a list.

Vertical lists are common in technical writing. A vertical list is in column format, and the elements may be indexed with numbers or marked with "bullets" such as "•." Usually you should omit terminal punctuation, whether commas or semicolons, in a vertical list unless it is a list of sentences.

All lists of both types must be "rhetorically parallel," meaning that the elements should be logically and grammatically similar such as all nouns in a list of the components of a system or all imperative sentences as the steps in a procedure. Rhetorical parallelism is grammatical similarity used to emphasize conceptual, and usually physical, similarity.

The Bullets and Numbering utility available on the standard toolbar and under the Format item of the main menu is especially helpful in composing a single column vertical list such as the following instructions for using this utility.

1. Begin the list by hitting the Enter key, to start a new line, and clicking on either the "Bullets" or the "Numbering" icon.
2. Every following new paragraph will be preceded by an index number or letter.
3. To change the index style use the Format, "Bullets and Numbering" menu items to open the "Bullets and Numbering" box.
4. It may be desirable to indent the right margin of the list items. This technique was used here to emphasize that the list belongs within the ongoing paragraph. Use the Format, Paragraph menu selections to open the Paragraph box.
5. Stop a list, after hitting Enter to end the last item in the list, by clicking the "Bullets and Numbering" icon off.

Obviously this utility makes a single column list almost automatic. Refining the paragraph format to increase the right margin is the only slightly sophisticated step.

The Table utility in MS Word is handy for creating multiple column lists. Multiple columns are helpful for lists of brief items to avoid a very long narrow list. Omit any bordering lines in or around such a list. One example follows, and others have been used throughout this manual. An example of a vertical list that was easy to construct using the table utility is this following list of suggested steps in effective lab preparation, conduct, and reporting.

1.	Preview lab manual and lecture notes.	7.	Process, analyze, and evaluate data.
2.	Attend preparatory lecture and preview.	8.	Outline report in notebook.
3.	Complete any pre-lab assignment.	9.	Draft report, emphasize exhibits.
4.	Arrive at lab promptly.	10.	Proofread and edit report.
5.	Conduct lab with attention to details.	11.	Use editing guidelines as final checklist.
6.	Gather data, make notebook entries.	12.	Submit complete report on time.

This example illustrates how a complex paragraph or section, burdened by minutia, can often be simplified and enhanced by moving the details to a list. The table utility made it easy to create the two column list. Text was merely entered into the cells. The left flush paragraph style, not right justified, is preferable for such a list. The gridlines are not printed in such a list. To format the gridlines use the Format, Borders and Shading menu item.

It is probably best not to end a paragraph or a section with the last item of a list. Such a termination is much too abrupt of a halt. Use a closing sentence at least, both as a transition and as a signal that the list is complete. In every case, be sure that a list is embedded in a paragraph with at least one preceding sentence to introduce and identify the list and at least one following sentence to complete the paragraph.

A list allows a complex paragraph or section, burdened by minutia, to be simplified and improved in organization by itemizing the details in a list. Good applications for lists, at least in informal reports, are steps in a procedure or components in an apparatus. Maintain rhetorical parallelism in all such lists. For examples, the steps can be imperative sentences, and the components can be noun phrases. Even in formal reports a list is often used in the closure to itemize the significant findings and conclusions. Obviously lists can be useful tools in efficient technical writing.

Summary Table

Table 13.1 summarizes the guidelines in this section and will be used as a guide to grading lists in undergraduate lab reports.

Table 13.1. Guide to Preparing Lists Used in Experimental Engineering Reports

Lists used when appropriate and helpful to simplify and enhance presentation.	List not used for incommensurate items.
Run-in list has no extraneous punctuation.	Vertical list can have bullets, numbers, letters, or no indices as appropriate.
Use multiple column list for brief items.	Optional right indention of vertical list is helpful.
List must be rhetorically parallel.	Consistent capitalization system used in all lists, usually initial capital only.

CHAPTER 2.14

NOTEBOOK POLICIES

Introduction

The research notebook is an important tool for organizing and preserving your data, ideas, and observations. Today we rely routinely on computer files, the contents and dates of which are easily altered; and this reliance may have actually made the research notebook even more important in research and scholarship and especially in significant legal proceedings. The notebook exists as a tangible record of priority in patent proceedings or as documentary evidence of good practice in cases of research integrity or product or professional liability actions. It should be brought to, and used in, every lecture and lab session.

Notebook Media Requirements

In graduate research or professional practice, the research notebook should be permanently bound (*i.e.*, pages sewn into the spine) with machine printed page numbers. For convenience and economy, undergraduates may be allowed to use an ordinary 8.5×11 inches spiral bound notebook in the undergraduate laboratories. Clearly identify the notebook with your name and affiliation (*i.e.*, your particular course) on the cover. Consecutively number by hand all pages you intend to use. Number the front and back pages if you write on both sides of a sheet. Do not number the entire notebook; save the balance for your next laboratory course.

Permanence is critical, so use **ink** not pencil. Line through ~~errors~~; do not obliterate them. The notebook is an important professional tool, so develop good practices now.

Leave no large blank blocks or gaps in the notebook. "Z-out" any unused blocks. These blocks create the suspicion that they have been left as opportunities to later enter fraudulent research observations or ideas for inventions.

If exhibits like graphs or printouts from instruments are inserted into the spreadsheet, they should be permanently pasted into the notebook. If you use tape instead of permanent paste to secure exhibits, initial over the edge to preclude tampering.

Special Requirements for Coursework

Some lab faculty usually impose a few special requirements to facilitate instruction in the areas of lecture notes, demonstration notes, and outlines. It may be required or recommended to use the notebook for lecture notes that do not fit into space reserved in the printed notes. In particular, include any special instructions from the instructor. Advance equipment demonstrations are sometimes conducted. These demonstrations are part of the lab experience. You should record pertinent aspects of the demonstration including a schematic diagram of the apparatus, a brief description of the principle of operation, and any especially significant quantitative information. In some courses, each student is required to prepare a report outline or list the objectives of each experiment in the research notebook.

Recording General Information

Each student must record all general information for every laboratory in his or her lab notebook. Examples follow: room T, atmospheric P, free stream velocity, heat exchanger configuration, pump speeds, and so on. Also identify all important instrumentation used including model and serial numbers if available. Routinely identify by name, date, and size the files that hold extensive computer based data. As always, date every entry.

Recording Specific Data

When the data set is brief, every student should record it all. For longer data sets record at least a representative sample. In either case, be sure to include the units of measurement. For example, the vapor pressure experiment illustrated in the attached example page below has only four data points. Every student should record all the temperature and pressure data for such a concise experiment. However, it should not be necessary for every student to record all the data for a more complex experiment. For example, the LDV traverse illustrated in Figure 8.1 in Chapter 2.8 has 34 pairs of data. For such longer data sets, record at least some sample data in the notebook both for practice and for later reference (*e.g.*, to later verify units of measurement).

Long data sets should be in your spreadsheet disk file with a copy distributed to every group member. It is good practice to print out a condensed version of the data block(s) and paste this exhibit into your notebook. Since a graph is a valuable aid to insight and analysis, consider pasting a miniature graph of the data into the notebook as well. Transparent tape is acceptable for course work, but in professional practice such exhibits should be permanently pasted into a notebook. If you are allowed to use tape instead of permanent paste, initial over the edge to preclude tampering. Never merely staple computer output or similar exhibits into the notebook. The resulting compilation is an invaluable resource, especially for a long duration research project. However, to save your time, inserting these exhibits is recommended but is not required in all undergrad lab courses.

Special Requirements for Patents

Information and data relating to conceptions that may be patented require special consideration. At a minimum, such entries should be read and signed by at least one knowledgeable and reliable witness.

Closure

In summary, follow a few simple guidelines. Always number the pages to be used. Date every entry. Every individual should record the general data (*e.g.*, room temperature and ambient pressure) for each lab. Every group member should record important observations and at least a representative sample of the routine data along with the verified units of measurement. List the important instrumentation and equipment. A generic description along with commercial identification (*e.g.*, manufacturer, model, and serial number) is best. Routinely identify by name, date, and size the files that hold extensive computer based data. In professional practice, permanently pasting a printout of the data into the notebook is advisable. For large data sets, consider a shrunken version of the printout. Since a graph is a valuable aid, consider including a miniature graph in the notebook. The notebook is an important professional tool so develop good practices now.

An example notebook page, prepared for a vapor pressure experiment, follows this section. Note that this page exemplifies the three most important features of a research notebook: the pages are numbered, the appropriate notes and data are recorded and dated, and the notebook represents a continuous and contiguous record of your activities. Table 14.1 itemizes the minimal guidelines for effective maintenance of the research notebook.

Table 14.1. Guide to Maintaining Experimental Engineering Research Notebook

Number all pages to be used.	Date all entries.
Use ink for all entries.	Z-out any large areas left blank.
Maintain a continuous and contiguous record.	Record all pertinent information, data, and observations.
Identify all important experimental equipment.	Record all general data for every experiment.
Record at least a sample of detailed data.	Optionally, paste in a spreadsheet block or other listing of computer based data.
Optionally, include a miniature graph of the data.	

SAMPLE PAGE FROM EXPERIMENTAL NOTEBOOK

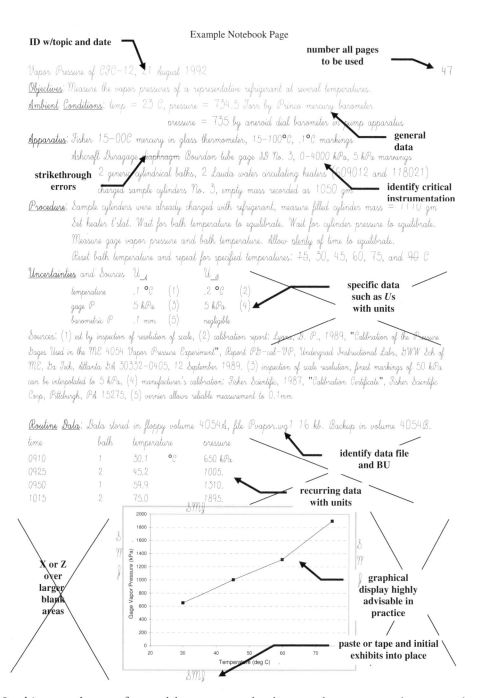

Example Notebook Page

ID w/topic and date

number all pages to be used

47

Vapor Pressure of CFC-12, 21 August 1992

Objectives: Measure the vapor pressures of a representative refrigerant at several temperatures.

Ambient Conditions: temp = 23 C, pressure = 734.5 Torr by Princo mercury barometer

pressure = 735 by aneroid dial barometer on pump apparatus

general data

Apparatus: Fisher 15-00C mercury in glass thermometer, 15-100°C, .1°C markings

Ashcroft Duragage diaphragm Bourdon tube gage ID No. 3, 0-4000 kPa, 5 kPa markings

strikethrough errors

2 generic cylindrical baths, 2 Lauda water circulating heaters (609012 and 118021)

charged sample cylinders No. 3, empty mass recorded as 1050 gm

identify critical instrumentation

Procedure: Sample cylinders were already charged with refrigerant, measure filled cylinder mass = 1110 gm

Set heater t'stat. Wait for bath temperature to equilibrate. Wait for cylinder pressure to equilibrate.

Measure gage vapor pressure and bath temperature. Allow plenty of time to equilibrate.

Reset bath temperature and repeat for specified temperatures: 15, 30, 45, 60, 75, and 90 C

Uncertainties and Sources U_A U_B

specific data such as *U*s with units

temperature	*.1 °C*	*(1)*	*.2 °C*	*(2)*
gage P	*5 kPa*	*(3)*	*5 kPa*	*(4)*
barometric P	*.1 mm*	*(5)*	*negligible*	

Sources: (1) est by inspection of resolution of scale, (2) calibration report: Lyons, D. P., 1989, "Calibration of the Pressure Gages Used in the ME 4054 Vapor Pressure Experiment", Report PG-cal-VP, Undergrad Instructional Labs, GWW Sch of ME, Ga Tech, Atlanta GA 30332-0405, 12 September 1989. (3) inspection of scale resolution, finest markings of 50 kPa can be interpolated to 5 kPa, (4) manufacturer's calibration: Fisher Scientific, 1987, "Calibration Certificate", Fisher Scientific Corp, Pittsburgh, PA 15275, (5) vernier allows reliable measurement to 0.1mm

Routine Data: Data stored in floppy volume 4054A, file Pvapor.wg1 16 kb. Backup in volume 4054B.

time	*bath*	*temperature*	*pressure*
0910	*1*	*30.1 °C*	*650 kPa*
0925	*2*	*45.2*	*1005.*
0950	*1*	*59.9*	*1310.*
1015	*2*	*75.0*	*1895.*

identify data file and BU

recurring data with units

X or Z over larger blank areas

graphical display highly advisable in practice

paste or tape and initial exhibits into place

In this example page from a laboratory notebook, note that page number, general information, and representative data are entered.

Chapter **2.15**

Computer Policies and Guidelines

Introduction

It is anticipated that students will at least begin—and in many cases finish—their laboratory report writing assignments on computers provided in a laboratory cluster or on the separate computers distributed through the lab areas. These computers should have adequate hardware and software for all of your anticipated needs. It is likely that many reports will be finished on your own personal computer. Students should note that an important component of Excel, the regression package, and an important component of Word, the equation editor, are skipped in typical installations. Please see the pertinent following subsections for information about adding these critical components.

To enhance the efficiency and effectiveness of your work, please also review the guidelines on some specific popular software presented in the last few subsections.

Public Computer Usage

In an instructional cluster, you may use a designated directory for temporary storage of your document and data files; however, you should always copy your files to your own media, such as a flash drive, and back this up to your home computer for permanent storage. Assume that the temporary directories will be periodically purged. Consider the following guidelines when using any public cluster computers.

1. Use your own flash drive—or other data storage medium—during your data processing. Back up your work often. Students are responsible for loss of data due to failure to preserve back up copies.
2. No unauthorized programs are permitted to be run on the computers located in public clusters.
3. Do not alter the program preferences or options or other software settings on any of the applications such as the document processor or spreadsheet programs. Please leave the software in the same configuration as you found it to avoid confusing or delaying the next user.
4. It is unlawful and dishonest to duplicate copyrighted material so do not copy any applications software from the public computers.

5. If a system malfunctions, students should please inform the user assistant, a lab or cluster teaching assistant, the lab supervisor, or a laboratory faculty member as soon as possible.

6. The hard disks are for permanent applications programs only. Do not alter, copy, or change any of the program files on the hard disks. Students may use the designated directory for temporary storage of files, but do not rely on the hard disks for permanent storage of any personal files.

These guidelines should help you to avoid the common problems encountered with public computers.

Other Laboratory Computers

The same guidelines typically apply for computers other than public cluster computers such as computers used with or integrated into experiments with the one exception that no personal files should ever be installed on such computers. Also, please take special care to scan your diskettes before installing them in such lab computers. Virus contamination that appears during an experiment would be especially destructive.

Software Usage

Most users will have little difficulty with Excel, Quattro-Pro, or MS Word except possibly for missing program components on personal systems. With some familiarity gained by direct experience, you can deal with most problems or needs by relying on your intuition and/or the online help utilities, tempered with some patience of course. Nevertheless, a few obscure or disintuitive features persist in every application. A few of these quirks are identified, and some procedures to resolve or avoid related problems are outlined in the following subsections.

Some Obscure Features in Excel

This program is both widely available and generally easy to use. For experimental engineering students, the most common significant problem is a missing regression package. This occasional, but significant, problem and a few minor issues are discussed in the following list:

1. *Regression analysis component missing*
   The standard installation procedure does not load the analysis tool package, leaving Excel unable to perform simple regression analysis or other related advanced numerical tasks. Normally the regression utility is reached by clicking the Tools option on the main menu and then clicking the Data Analysis option to open the Data Analysis dialogue box. The regression option should be listed as an option in this dialogue box. If the Data Analysis option does not appear on the Tools menu, then the data analysis package must be added.

On the Tools menu, click instead on the option called Add-Ins. The Add-Ins dialogue box will appear. Use this box to load the Data Analysis ToolPak. The Excel or Office installation CD will be needed.

2. *Avoiding the line chart*
    Excel, like other spreadsheet programs, has extensive graphics capacity. For engineering graphs the "XY" graph type is almost always used, specifically avoiding the "Line" type, which is only a business oriented presentation, not a mathematical plot. It's best to define a convenient default graph. Otherwise use the graph "wizard" to generate a graph, and use the graphics editing functions to modify the graph to suit your purposes.

3. *Inserting graphs into reports*
    A graph may be printed from the spreadsheet and attached to the report, but this practice is discouraged except when preparing a separately printed graph for a projected visual presentation. It is preferable to insert the graph as an object into the text. To copy and paste a digital version directly into the document use the "Paste Special, Picture" function in MS Word.

4. *Graph area prints solid black*
    An occasional problem occurs when the plot area in an Excel graph is needlessly defined to be transparent with no area fill color. This specification causes some printers, such as the Lexmark, to print a solid black graph area. This problem cannot be duplicated on an HP printer. Avoid this problem by merely leaving the plot area color on Automatic, which gives a white default area color that is not misinterpreted by the printer.

5. *Using the shortcut for address types*
    When editing formulas in either Excel or in Quattro-Pro, use the F4 key to cycle among the four possibilities of relative and absolute cell addresses.

Some Obscure Features in MS Word

Most operations and options in MS Word are largely intuitive, but a few options important in technical applications are rather well hidden and are listed below. The third item in the list addresses the important problem when the equation editor is not loaded.

1. *Starting the equation editor from menu bar*
    To insert an equation using the menu bar, click on Insert and select Object. The Object dialogue box should open with Microsoft Equation 3.0 as one option.

You can now edit the equation. All of the usual symbols for mathematical operations and structures and the Greek alphabet are available from pull-down menus in Equation Editor. For specific guidance see the section on equations in this manual.

2. *Missing equation editor*

   When the equation editor is missing, then it will not appear as an option in the Insert Object dialogue box. It is then necessary to install this missing component. The following instructions should work for users of the Microsoft Office suite of programs. The MS Word or Office program disk will be needed. The procedure for a separate installation of the MS Word program will be similar but slightly different.

   (a) Close all programs and return to the desktop. Then select Settings from the Start menu.
   (b) From the Setting box select Control Panel.
   (c) From the Control Panel box select Add/Remove Programs.
   (d) From the Add/Remove Programs box select MS Office, or Word if installed separately, and click the Add/Remove button.
   (e) Insert the MS Office, or Word, compact disk into the CD drive.
   (f) From the MS Office Setup box select Add/Remove.
   (g) From the MS Office Maintenance box select Office Tools and click the change option button. Note especially that the Office Tools option is selected, not the MS Word option.
   (h) From the Office Tools box select Equation Editor to be installed. Click the continue button to install the missing component.

3. *Using equation editor icon*

   If this icon, which is the radical operator enclosing an alpha or $\sqrt{\alpha}$, is present, start the equation editor with it.

4. *Missing equation icon*

   If the equation icon, $\sqrt{\alpha}$, is missing, add it to the Standard toolbar from the Insert toolbar or from the objects available for the Insert toolbar using the typical procedure. First on the main toolbar select the View item to pull down that menu. There select the Toolbox item to open that submenu. There select Customize to open the Customize dialogue box illustrated in Figure 15.1. Select Commands and then the Insert category. Find the equation icon under the Insert category and click and drag it to a position on the Standard toolbar. This process should give you the equation icon readily available for future use. In case of problems, refer to the Help utility for assistance.

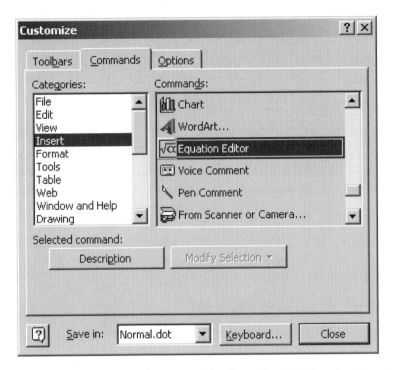

Figure 15.1. The Customize Dialogue Box Configured to Display the Equation Icon

5. *Other icons*

   Some considerable time will be saved by adding three other icons to the Standard toolbar. These are the superscript icon, x^2, and the subscript icon, x_2. These obviously handy icons are available from the Format menu. Another handy icon is the Insert Symbol icon, Ω. This icon is handy for entering the numerous symbols encountered in the text of technical reports that are discussed in the next paragraph.

7. *Symbols and special characters*

   Many characters and operators are needed from the symbols font or character list. Greek letters and the usual math symbols and operators are common examples. Some important but less obvious examples are the degree symbol, °, the multiplication dot, ·, and the copyright mark, ©. Other examples are the forward arrow, →, that may be needed in chemistry text and the back arrow, ←, that may be needed as the replacement operator in computer code or pseudo-code. When the negative sign is needed in text use the symbol, –, not the keyboard dash, -, symbol from the standard keyboard. Obviously, it is convenient to have the Insert Symbol icon on the toolbar to facilitate these entries.

8. *Centering tables*

When editing tables in WORD 97 and later versions, position the cursor on the table and enter the following command

Table, Select (entire) **Table**

Now use the centering icon on the standard toolbar. Alternatively, drag the selection arrow just to the right of the table. This action should also select the entire table not just its contents. If only the contents of the table, not the entire table, is selected then the text will be centered within the cells.

9. *Drawing grid lines*

Outside border and internal grid lines may be drawn over a table by selecting the table and using the Format, Borders and Shading menu. Select the desired border style from the dialogue box. The preset "Grid" style is handy and appropriate for many applications.

10. *Inserting a Quattro-Pro graph into written report*

A Quattro-Pro graph may be copied and pasted into a Word document using the Windows clipboard, or a bitmap file can be generated and then inserted into the report. Note that bitmap files are typically very large. Once inserted either image cannot be edited extensively but only repositioned or resized as a "Picture" or single graphic element. In either case a quick trial will demonstrate that the "Float over text" option should be deactivated before pasting the image into the document.

CHAPTER 2.16

SIGNIFICANT DIGITS AND UNCERTAINTY

Basic Concepts and Types of Uncertainty

Uncertainty is not the result of sloppy measurement. Instead, every measurement has some uncertainty, and it is sloppy procedure to ignore the uncertainty. In general the best available expected value is used as the measurement, and it should be known or at least estimated to lie within some confidence interval. The half-width of the confidence interval is the uncertainty. By convention, the confidence interval is usually taken to be the 95% confidence interval, meaning that there is good reason to be 95% sure that the actual value lies within the interval. The uncertainty value that defines the limits of the confidence interval is formally called the Expanded Uncertainty, and it is symbolized by U. Ideally, a measurement, m, would best be expressed as

$$m \pm U \qquad (16.1)$$

The uncertainty and the measurement are illustrated in Figure 16.1. In the figure, the measured value is taken, as usual, to be the middle point in the confidence interval. By interpolation, the measured value is 6.3 units, and by observation or analysis of the measurement apparatus and process, the Expanded Uncertainty is taken to be 0.1 unit, so the measurement should be expressed as 6.3 ± 0.1 units.

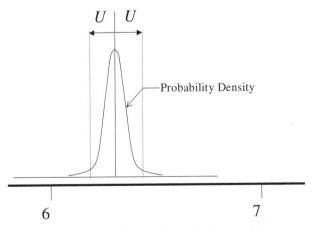

Figure 16.1. Schematic Illustration of a Generic Measurement

Uncertainty results from two general classes of errors, random fluctuations and systematic error or bias. Uncertainty due to random errors can usually be statistically addressed and is formally called Uncertainty A (U_A). The older informal term analogous to this uncertainty is *imprecision*, meaning the lack of perfect reproducibility in the measurements. Systematic uncertainty due to bias also always exists. It must usually be assessed by an analysis of the entire measurement system, and it is formally called Uncertainty B (U_B). The older informal term analogous to this uncertainty is *inaccuracy*. The overall effect of both types of uncertainty is called *Combined Uncertainty*. The two types of uncertainty are illustrated in Figure 16.2.

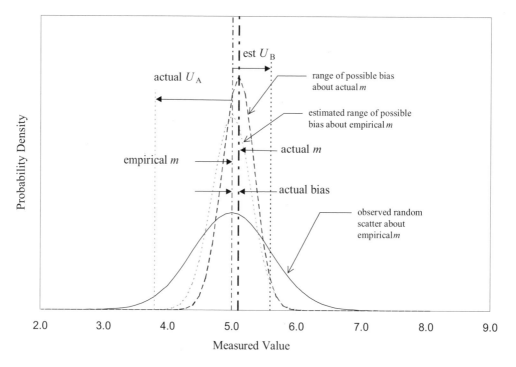

Figure 16.2. Idealized Schematic of the Types of Uncertainty and the Associated Distributions

The uncertainties and their related distributions are illustrated in Figure 16.2. The figure shows an empirical measured value at 5.0 units. The broad bell-shaped curve shows the considerable random scatter associated with this measurement. Presumably, a number of measurements have been made, and the average and the scatter in the distribution have been computed. The average gives the empirical measurement m, and the scatter in the distribution is used to assess the actual U_A. Also note that the actual underlying value of m is somehow known to be at 5.1 units. In principle, this underlying actual value could never be known experimentally. Assume that the possible systematic bias in the data has been estimated. For example, assume that, because of the possible systematic error in the measurement system, empirical values for the averaged measurement could range from 4.6 to 5.6 units. The half width of this actual range of possible bias is 0.5 units, which is the U_B for this measurement. Neither the actual bias nor the actual possible range in bias can be determined. The actual bias itself cannot even be estimated. If it could be, the investigator would merely correct for the bias. However, the range of the possible bias can be estimated by techniques presented below. Working knowledge of the uncertainties is necessary to conduct and describe experimental work, so this section will review techniques for determining the U_A and U_B.

Some estimate of the uncertainty is always necessary for any measurement to be useful in experimental engineering or design. For example, the uncertainty is needed in order to determine if the measurement agrees with alternative experimental data or a theoretical prediction, and obviously an uncertainty is needed in order to assign a reasonable margin of safety in design. In practice, many engineers avoid making an explicit assertion of the uncertainty, but one can be sure that, in most cases, competent engineers have an implicit, even if private, evaluation of the uncertainty in mind. Indeed, at least an order of magnitude estimate of the uncertainty is necessary just to identify the significant digits in the data, where the significant digits are the digits that actually convey useful information about the measurement.

The digit in the smallest place that represents any useful information is called the least significant digit (LSD). It is necessary to know the uncertainty in order to identify the LSD, because the LSD in a measurement or other empirical value should correspond with the LSD in the corresponding uncertainty. For example it is proper to write $3.24 \pm .18$ kg but not $3.242 \pm .18$ kg. In technical writing, it is important to never report meaningless trailing digits that contain no information. To avoid such errors with significant digits, the writer should understand the theory and practice of uncertainty and significant digits, so a brief synopsis of the relationship between the two concepts is presented in this chapter. For quick review and routine guidance, a summary of the accepted practical rules on significant digits is presented in a following section.

Types of Measurements

With respect to uncertainty analysis, two types of measurements are encountered in the academic or research laboratory and in practice. When a measurement is read from an instrument in the units in which the instrument is graduated, the measurement is direct. For example, the width and thickness of a small metal bar can be measured directly with a micrometer caliper, and the measurements can be read directly from the barrel of the micrometer. The cross sectional area or moment of area of the beam can then be calculated, and the result of the calculation is an indirect measurement.

Types of Experiments

When the experimental conditions are held fixed and the necessary direct measurements are made and the associated indirect measurements are calculated, the results are called *single point measurements* by at least some authors. Obviously, this term could be confused with the practical design of an apparatus with a single measurement location, so to avoid confusion, such a measurement at fixed conditions will be called herein a *single-point experiment*. For example, the temperature can be held constant and the mass of liquid in a pychnometer, which is a vessel of well-known volume, can be measured. Then, the density can be calculated and reported as an indirect measurement.

By analogy, when the controlled variable is changed over some deliberate range and a set of measurements is made, the result could be called a *multiple point measurement*. This term is not used much in the literature and would be easily confused with the practical description of an apparatus with multiple measurement locations. For convenience, the specific and descriptive term *multiple point experiment* will be used in this text. The direct measurements are not much different in a multiple point experiment, but distinctive indirect measurements can be made particularly as the result of regression analysis. While other more sophisticated multiple point measurements exist, the most common are regression models, and these are the only multiple point measurements addressed in this text.

Overview and How to Use This Chapter

In the following sections, after the necessary definitions are stated in logical order, applications of both Uncertainty A and Uncertainty B to direct and indirect measurements in both single point and multiple point experiments will be considered. Note that there are two types of uncertainty, A and B, there are two types of measurements, indirect and direct, and two types of experiments, single and multiple point. There would seem to be eight combinations to consider, but actually the practical combinations are fewer. First, there are actually no multiple point direct measurements since all such results are calculations and inherently indirect measurements. Also, there is really nothing distinctive in the treatment of the types of uncertainty with respect to single point indirect models, which further simplifies the cases.

The knowledgeable reader can skip or defer the background material and refer directly to the tables where the rules or guidelines on significant digits and uncertainties are summarized.

Definitions

There are a few specialized terms relating to uncertainty and some everyday words that have specific meanings in this field. The pertinent definitions are listed as follows:

The *significant digits* are the digits that actually contain useful quantitative information.

The *Expanded Uncertainty, U,* is the 95% confidence limit for a measurement. This value is often called just the "uncertainty."

The *Standard Uncertainty, u,* is a statistic such as a Sample Standard Deviation used to compute the *U* according to the general formula

$$U = k_C\, u \tag{16.2}$$

The *Coverage Factor* is the multiplier k_C in the preceding formula. It gives the half width of the confidence interval in terms of *u*. For a very large data base governed by the bell-shaped normal distribution, *u* is the Standard Deviation; and 2.0 is the statistically correct value for k_C. For a small sample of data, *u* is the Sample Standard Deviation, and the value of k_C varies with the size of the sample. As discussed later, the value to be used for k_C for small samples should be determined from the t-distribution. Values for k_C are tabulated in the appendix to this section.

The *Uncertainty A* is one of the two components of uncertainty. In common terms, U_A results from random variation in the data. In principle the U_A is defined operationally as the uncertainty that can be evaluated by statistical analysis of repeated or varying measurements. As discussed later, the U_A can often be estimated in practice for single measurements by considering the analog graduation or the digital resolution of the instrument or by observing the actual fluctuation or variation in the data. In older publications, the concept analogous to this component of uncertainty was referred to as the lack of "precision," meaning the random error or lack of perfect repeatability in a set of measurements.

The *Uncertainty B* results from systematic error or possible bias in the data. In principle, estimation of the U_B requires analysis of the measurement system. Analysis of repeated measurements will reveal nothing about the U_B. The U_B from calibration is typically determined during the calibration process and should be reported for use with the calibrated instrument. A rough value for the U_B for an uncalibrated instrument can typically be estimated from accumulated general experience with similar instruments. Usually, the general method of Error Propagation Analysis combined with appropriate physical analysis is used to determine the U_B. In older treatments, the analogous form of this uncertainty was referred to as the systematic error or imperfect "accuracy."

The *Combined Uncertainty* or U_C is the unified effect of the Uncertainty A and the Uncertainty B. The rule for combining uncertainties is one topic reviewed in a later section. Since the two components are surely independent sources of error, the U_C is given by

$$U_C^2 = U_A^2 + U_B^2 \qquad (16.3)$$

The *resolution* is the finest possible measurement. Typically, it is the smallest digit that is stable enough to be read from a digital display, or it is the result of interpolation from an analog scale or dial.

The *graduation* is the finest increment of markings on an analog instrument. The analog graduation is typically an order of magnitude larger than the resolution. For example it should be possible to interpolate most mercury in glass lab thermometers graduated at 1 degree intervals to *ca.* 0.1 or 0.2 degree.

A *Calibration* is the process or the result of comparing an instrument with a standard. For examples, a thermocouple can be calibrated by comparison with a standard resistance temperature detector (SRTD), and a pressure gage can be calibrated by comparison with a dead weight tester.

The *Calibration Function* gives the corrected value as a function of the measured value. An example of calibrating a pressure gage is

$$P_{corr} = c + b\,P_{mes} \pm U_{C,\,CAL} \qquad (16.4)$$

The $U_{C,CAL}$ is the combined uncertainty of calibration. Hopefully, this uncertainty will be relatively small. In the ultimate application the $U_{C,CAL}$ is fixed, so it must be taken to be a source of possible bias with respect to the actual measurements.

A *Direct Measurement* is taken from the scale or display of the instrument in the units in which the instrument is graduated. Reading 124 kPa from the dial of a Bourdon tube pressure gage is a direct measurement.

An *Indirect Measurement* is calculated from other more direct measurements, which may be direct measurements or intermediate indirect measurements. If the 124 kPa direct measurement were adjusted using a 1.05 calibration factor, the corrected pressure of 130 kPa would be an indirect measurement.

The *Measurement Formula* gives the indirect measurement, y, as a function of one or several more direct measurements, so

$$y = y\left(x_1, x_2, \cdots x_n\right) \qquad (16.5)$$

where the x_i are the more direct measurements.

The *3-to-30 Rule* requires either one or two digits for U adjusted so the value of U lies between 3 and 30 in units of the least significant digit. This rule is promoted by some authorities on experimental practice and technical writing, but is not nearly universally adopted.

The *Standard Deviation* (SD) is a measure of the dispersion in a large sample of data with respect to its mean. It is computed by the formula,

$$SD^2 = \frac{\displaystyle\sum_{i=1}^{N}\left(x_i - \mu\right)^2}{N} \qquad (16.6)$$

where N is the number of data in the sample, and μ is the mean or central value of the infinite population from which the large sample is taken. When the dispersion in a population is generated by small random fluctuations from the mean, the statistics of the population are governed by the classical Gaussian or Standard Normal Distribution, which is the familiar bell shaped distribution. In this distribution, about 68% of the data lie within 1 SD of the mean, about 95% lie within 2 SD of the mean, and more than 99% of the data lie within 3 SD of the mean.

The *Sample Standard Deviation* (SSD) is a measure of the dispersion in a finite sample of data. It is computed by the formula,

$$\text{SSD}^2 = \frac{\sum_{i=1}^{N}\left(x_i - x_{\text{AVG}}\right)^2}{N-1} \tag{16.7}$$

where N is the number of data in the sample, and x_{AVG} is the usual arithmetic average. The SSD differs from the SD only by the reduction of one in the numerator. Note that the true mean of the hypothetical infinite population from which the sample was taken is unknown experimentally; therefore, the mean must be estimated by the arithmetic average. This estimate introduces possible bias into the estimate of the dispersion. For example, one large datum in the sample would make the average unrepresentatively large, and since the average now overestimates the mean, the dispersion would be underestimated. The numerator is appropriately reduced to compensate for this possible bias. Note that the SSD becomes essentially identical to the SD when the sample is reasonably large, say 30 or more data.

The *Standard Error of Estimate* is a measure of the scatter of data with respect to a regression model. The formula for this statistic is

$$\text{SEE} = \sqrt{\frac{\sum\left(y - y_{\text{est}}\right)^2}{DF}} \tag{16.8}$$

where y_{est} is the corresponding value of the regression model, and DF represents the index called the statistical *Degrees of Freedom*, which is the number of data less the number of parameters.

The common powers of ten notation is called *scientific notation*. Only the significant digits are written in scientific notation. For example 3.165×10^3 and 3.160×10^2 both have four significant digits.

The *order of magnitude* of a measurement is roughly the closest power of ten to the measurement. A reasonable quantitative definition is the integer that is closest to the common logarithm. For example 3.165×10^3 has common logarithm 3.500 and order of magnitude 4.

Both the application and the significance of these definitions are described in the following sections.

IDENTIFYING AND REPORTING SIGNIFICANT DIGITS
Overview and Background
Introduction

Sometimes the Combined Uncertainty is known, and it is the applicable uncertainty. At other times, only the Uncertainty A is known or is applicable. In any case, the uncertainty dictates the LSD. The least significant digit in a measurement must correspond with the least significant digit in the associated uncertainty. For example consider the measurement $3.24 \pm .18$ kg. Because of this intimate relationship between the LSD and the uncertainty, at least an order of magnitude estimate of the uncertainty is needed to identify the LSD. Assuming the uncertainty is known or has been estimated, the LSD can be identified. Then, the measured data and calculated data, called the direct and indirect measurements, can be properly reported. The first topic that must be considered is rather fundamental, finding the reasonable number of significant digits in the uncertainty itself. The next topic is then essentially straightforward, finding and reporting the significant digits in the associated measurement.

Number of digits in the expanded uncertainty U

Note that the first digit in the uncertainty corresponds to a digit in the measurement that is somewhat uncertain. Furthermore, the second digit corresponds to a measurement digit that is highly uncertain but not worthless. In contrast, a three-digit uncertainty, such as 2921 ± 242 mm, is ridiculous. Observe that the digit in the third place in the uncertainty corresponds here to the 1 in the measurement. This digit could, as in this extreme example, be barely one-thousandth of the value of the digit corresponding to the first place of the uncertainty, which is 900 in this case. This larger value is itself somewhat uncertain, so the digit that is almost one thousand times smaller must be useless junk. Consequently, an uncertainty must reasonably have no more than 2 significant digits, or the third digit would correspond to an entirely useless random value.

Unfortunately, no universal rule is available to define whether an uncertainty should have one or two digits. Indeed at least three situations or recommendations exist in practice and in the literature. A few authorities suggest only one digit (*e.g.*, $24.4 \pm .2$ kg). Examples of respected texts that promote this rule are Skoog (1969, pp. 50–51) and Massey (1986, pp. 91–92). This rule seems somewhat restrictive, but it is clearly appropriate in those cases where the resolution of the instrument limits the resolution of the uncertainty. In the example cited above, assume the mass of 24.4 kg were measured by a beam balance graduated at intervals of 1 kg that could be reliably interpolated to $\pm .2$ kg. In this case, a one-digit uncertainty with a magnitude at the limit of resolution is appropriate. Others suggest two digits (*e.g.*, $3.24 \pm .34$ kg) in the uncertainty. This two-digit rule is advocated in well-established texts such as Palmer (1912, pp. 58–59) and is exemplified in authoritative references such as the guide published by the NIST (Taylor and Mohr 2002). This rule seems appropriate particularly if the Uncertainty A is determined by statistical analysis of repeated measurements and/or the Uncertainty B is determined by Error Propagation Analysis. In these cases, the second digit should be available from the calculation, and it should be meaningful in application. Others suggest using either one or two digits adjusted according to the 3-to-30 rule that is discussed in the next subsection. All three of these rules are reasonable if applied appropriately. A recommended practical rule for general student use is suggested in the next paragraph.

Recommended rule

Since at least three similar but distinct rules for the number of digits in the U exist, a consensus or compromise rule is necessary for consistency in the undergraduate laboratory. The recommendation is to use the simple two-digit rule whenever a two-digit uncertainty can be calculated or estimated and to revert to the one-digit rule when necessary. Two-digit uncertainties should be feasible when sampling data and calculating the uncertainties or when using instruments calibrated to two digits of uncertainty. A one-digit uncertainty as a general rule may be undesirable; however, one digit is quite reasonable when the instrument is not calibrated or graduated more finely. Therefore, use a one-digit uncertainty when limited by resolution or calibration. If a scale ruled at 1.0 mm intervals were used to measure a length, a measurement of $28.6 \pm .2$ mm seems reasonable since careful visual interpolation should yield an uncertainty on the order of 0.1 mm. An interval smaller than 0.1 mm would surely be invisible, so the limit of resolution limits the number of digits. No one should reasonably object to a one-digit uncertainty such as $\pm .2$ mm in such a case. Indeed one-digit uncertainties should be the expected result from reading most analog instruments directly. This flexible rule should be suitable in most situations; however, some editors or instructors will reasonably prefer to use the alternative 3-to-30 rule described in the next section.

Background of the 3-to-30 rule

The objective of the now unknown original proponents of this rule was apparently to define a simple criterion for dropping the second digit when it is a relatively small fraction of the first digit. An alternative idea that gives background for this rule would be to drop the second digit when it amounts to less than say 5.0% of the first digit. By this rule, every two-digit uncertainty in the range from 90 to 100 would be rounded and then truncated to 9 or 10, and only 84 and 85 would be retained in the decade from 80 to 90. In contrast, every second non-zero digit would be kept in the range from 10 to 20 and only 29 would be truncated in the range from 21 to 30. While this proposal is logical, it is obviously quite complicated. The writer would need to remember or verify that 42 should be retained because truncation would cause a 5% error while 41 is small enough to be truncated. The traditional alternative is simpler: just require that the uncertainty must fall in the range from 3 to 30 in units of the LSD. This rule keeps intact the range from 11 to 28 where rounding and truncating would cause errors from 5% to 40%, while specifying one digit uncertainties in the range where rounding and truncating the second non-zero digit cause acceptable under or over estimates of only 1% to 14%. This rule is recommended in respected scientific texts such as Shoemaker (1996).

Details of the 3-to-30 rule

This rule seems to be conveyed more by tradition than by publication in the literature, but a reasonable interpretation and additional justification can be presented here. Begin with the smallest two decades of two digit uncertainties in the range from 10 to 30. All of these two digit uncertainties should be retained since as explained above the second non-zero digits amount to at least 3% or as much as 40% of the first digit. Next, it seems that all second digits for uncertainties in range 31 to 34 should be rounded to 30. This rounding retains the same place of the LSD with a possible underestimate of no more than 13% at worst. Uncertainties in the range from 35 to 94 should be rounded and then truncated to range from 4 to 9, because omitting the second digits in these cases results in acceptable under or over estimates of only 1% to 14%. The smallest error is rounding 91 to 90 and reporting 9 for only 1% error. The worst case is rounding 35 to 40 and truncating to 4 for a marginally acceptable 14% error. These errors are assumed to be insignificant fractions of an already uncertain digit, and the dropped digits should contain little or no useful information. Uncertainties in the range of 94 to 99 are rounded to 100 and then truncated to 10. This new uncertainty has two significant digits and would properly be written as $10. \times 10^2$ units to emphasize the new place of the LSD. Note that all second digits in the uncertainties from 10 to 30 are kept. As demonstrated above, this range was apparently selected to give a simple rule for retaining the second digits only in the range where most of the second digits are relatively large. In fact most of these second digits are important, since truncating to one digit in this range can lead to some substantial errors. Here the extreme case is rounding an uncertainty of ± 14 units to ± 10 units. This 40 % underestimate would be unreasonable, so the original uncertainty of 14 units is retained. The second digits that are kept have enough information to be valuable.

Examples of the 3-to-30 rule

By definition, this rule requires either one or two digits for U truncated if necessary so the value of U lies between 3 and 30 in units of the least significant digit. For example, measurements of $3.24 \pm .18$ kg or $3.2 \pm .4$ kg are acceptable. In the first case, the uncertainty is 18 units of the LSD that is in the hundredths place, and 18 is between 3 and 30. In the second case, the uncertainty is 4 units of the LSD that is in the tenths place, and 4 is between 3 and 30. A measurement of $3.24 \pm .58$ kg is not acceptable under the 3 to 30 rule. This U should be rounded to 0.6 kg and the measurement reported as $3.2 \pm .6$ kg.

Applying the alternative 3-to-30 rule

This rule is blessed by tradition and is popular with some instructors and editors. More importantly, it is a simple way of restricting the uncertainties to a reasonable range where the second digit is substantial. Since this rule is entirely reasonable and can be applied with minimal extra effort, no one should complain if two-digit uncertainties are constrained to fall within these limits. The only reasonable objections should be a strong reluctance to upgrading inherently one-digit uncertainties of 1 or 2 units to 10 or 20 units one decade smaller. If the experimental reality does not support this change, the experimenter should properly object and use the one-digit values.

  With the number of digits in the uncertainty understood, it is easy to address the reporting of significant digits in the measurement.

Reporting Significant Digits

In this section, the rules for reporting significant digits when the place of the LSD is already known are reviewed. Some rough idea of the uncertainty is necessary for the place of the LSD to be identified. Fortunately, the very nature of experimental work dictates that the uncertainty can be evaluated. For example, no one really hesitates to write down the value of a simple direct measurement usually in the correct number of significant digits because the number of significant digits and the place of the LSD are inherent in a direct measurement. Sometimes the appropriate place of the LSD is not obvious, and then the nature of the direct measurement or the operations in the calculations that generate the indirect measurement must be analyzed to determine the uncertainty or at least the place of the LSD in the uncertainty. This analysis and some short cut rules are addressed in the next subsection. Once the uncertainty is known at least approximately, researchers and practitioners should take care to report only significant digits in formal work such as tables and text. The conventions for reporting measurements so that the significant digits are obvious are explained in the next section.

General rule

The general rule for reporting significant digits is simple and logical: all non-zero digits and all zeros that are not mere place keepers must be significant digits. Follow this rule to avoid asserting significance to digits that are only insignificant junk, and never include trailing insignificant digits in the formal text or tables of reports. To make the general rule easier to follow, some specific cases are addressed in the next section. If the writer takes care in following the general guidance, it will be routine to report numbers so that the significant digits are correct and obvious to the reader.

General guidance and examples

The first part of the general rule is that all non-zero numbers anywhere must be significant digits as in 3.1416 mm, which must have 5 significant digits with the LSD in the ten-thousandths place. Zeros require special attention so that zeros that are merely place keepers are not erroneously interpreted as significant digits. Most important, all zeros left of the decimal and right of a significant digit must be significant digits, so 310. mm has three significant digits, but 310 mm should have only two significant digits. Trailing zeros are usually only place keepers like the 0 in the ones place in 310 mm. Report writers must be careful about the occasional trailing ambiguous significant zero or zeros. For example 3100 mm would be usually interpreted as having only two significant digits, the non-zero 3 and 1. However, if the uncertainty were 150 mm, then the measurement would actually have three significant digits in 3100 ± 150 mm. In this case, the zero in the tens place is a significant zero. The measurement 3100 mm would never be interpreted as having four significant digits since it would then be written as 3100. mm. Unambiguously reporting some significant zeros may require special attention as reviewed in the next paragraph. Zeros right of a decimal and left of a SD, which would necessarily be non-zero, are always merely place keepers. An example is .0031 mm, which has two significant digits. Some editors require that a decimal fraction be written with a leading zero as in 0.0031 mm. This leading zero is not to the right of a significant digit. It is completely redundant and is unnecessary even as a place keeper, so the measurement still has two significant digits. Finally, zeros right of the decimal and right of any other significant digit are not mere place keepers and must be significant digits. For example, .00310 mm has three significant digits since the trailing zero is to the right of the decimal and to the right of at least one non-zero digit; therefore, this trailing zero must be significant. More subtly, the measurement 2.0 mm must have two significant digits, since the zero is right of the decimal and right of the other significant digit(s). Recall that all digits in scientific notation must be significant as in 3.10×10^3 mm. Since every digit is significant, the trailing zero in this scientific notation is significant. Insignificant digits are acceptable in practical working spreadsheets attached to narrowly circulated reports. Indeed it is practically impossible to display only significant digits in a working spreadsheet. In practice the efficient procedure is to do the calculations, then identify the LSD, and only report significant digits in the text and tables of a report.

Specific guidance on significant zeros

Zeros that are significant can sometimes be overlooked and create some confusion. When an ambiguous significant zero appears in a table, inspection of the analogous digits in other numbers in the same column is usually enough to identify the zero as significant. In text, the significant zero may be overlooked. To prevent this problem, consider writing the measurement along with the uncertainty as in 3100 ± 150 mm or consider using scientific notation and write 3.10×10^3 mm. For absolute clarity, consider using scientific notation even in a table.

Statistics

Statistics are a bit more complicated than other measurements and are much better addressed later in this section. No general rules on significant digits have been widely adopted or apparently even proposed for statistics, and the rigorous rule has not been recognized or much used in practice. Consequently, it is best to adopt a reasonable but arbitrary stopgap rule for student work. As explained and justified in the following section on statistics, maintain enough SD in statistics to calculate a percentage or to make an unambiguous ranking.

Spelling out numbers

Literary style guides typically propose a rule to spell out small numbers, say less than 100. This rule has little relevance to technical reporting. It is hardly ever reasonable to ever spell out a real number that represents any kind of continuous measurement since the digits and placement of the decimal are needed to identify the significant digits. It may be appropriate to spell out very small integers that are presumably known exactly, but even for integers, the more general rule of stressing the quantitative nature of all data argues in favor of using numerals. At the least, using numerals surely avoids translation of numerical data that could be critical. In summary, it may be appropriate to spell out small and presumably exact integers, but other numbers should be expressed in numerals. Indeed, since integers are fundamentally different from real numbers, it is sometimes appropriate to identify an integer value explicitly as in "The integral number of fringes is 14."

Conclusion

Remember that the significant digits are the digits that contain appreciable information and must therefore be reported and that digits smaller than the LSD are mere junk that should be dropped. As a mental cross check, recall that every digit is a significant digit in scientific notation. Elsewhere, except for place keepers, the numerals should only be significant digits. The general rule and the important special cases are illustrated in the Table 16.1. Identify and report significant digits correctly to avoid damage to your credibility by reporting an absurdly large number of digits as being significant.

Table 16.1. Examples of Identifying and Displaying Significant Digits (SD)
The least significant digit (LSD) in the example entry is underlined.

Example Entry	Number of SD	Interpretation[a]
3.141<u>6</u> mm	5	All non-zero digits must be SD.
31<u>0</u>. mm	3	All zeros left of the decimal and right of a SD must be SD.
<u>3</u>10 mm	2	Trailing zeros are usually only place keepers.
.003<u>1</u> mm 0.003<u>1</u> mm	2	Zeros right of the decimal and left of a non-zero SD are merely place keepers.
.0031<u>0</u> mm	3	Zeros right of the decimal and right of a non-zero SD must be SD.
25<u>3</u> mm 31<u>0</u> mm 57<u>4</u> mm	3	Significance of a trailing zero may be inferred in tabulated data.
3.10×10^3 mm	3	All digits in scientific notation must be SD.
$U = .02\underline{3}$	2	When possible, uncertainties should have 2 SD.
$R^2 = 0.9995$ alt $R^2 = 0.999\underline{1}$	4	Maintain enough SD in these statistics to permit an unambiguous ranking.[b]
$\alpha = .001\underline{9} = .1\underline{9}\%$	2	Maintain enough SD in this probability to allow an accurate percentage to be calculated.[b]
$\alpha = .049\underline{9} < .050\underline{0}$	3	Maintain enough SD in these probabilities to permit an unambiguous ranking.[b]

[a] Except for the statistics, the LSDs in this table have been identified by this general rule: All non-zero digits and all zeros that are not mere place keepers must be significant digits.

[b] In these statistics and statistical probabilities, enough SDs are maintained to calculate a percentage or make an unambiguous ranking.

Principles of Uncertainty

Obviously, the uncertainty of a measurement is a critical consideration in any experimental work. As mentioned above, uncertainty typically results from two general classes of errors, random fluctuations and systematic error or bias. Uncertainty due to random errors can usually be statistically addressed and is formally called Type A uncertainty. The older informal term analogous to this uncertainty is *imprecision*. Systematic uncertainty due to bias must usually be assessed by some other analysis and is formally called Type B uncertainty. The older informal term analogous to this uncertainty is *inaccuracy*. The basic principles of uncertainty will be reviewed in this section. In following sections, the principles will be applied to some important cases.

Recall that the range of minus one to plus one standard deviation (*i.e.*, an S.D. or σ) from the mean in the normal distribution includes 68% of the possibilities. The experimental uncertainty that is analogous to the standard deviation is called the Standard Uncertainty. The general symbol u is used for this value. The Standard Uncertainty is usually not quoted in reports. Instead, it is typical to report a larger range that completely spans the main body, typically 95%, of the possible values. This overall uncertainty is called the Expanded Uncertainty, and the symbol U is used for it. In general, the Expanded Uncertainty is computed by the trivial formula,

$$U = k_C\, u \tag{16.9}$$

where k_C is called the coverage factor. If the experimental data base is infinite or reasonably large and the variation in the data results from the random accumulation of small independent errors, then the normal distribution applies, and $k_C = 2.0$. In practice, a sample of 30 or even 20 data can be large enough to be considered "infinite." In more general and typical cases when the experimental sample size is small, a slightly more sophisticated analysis using small sample theory and the t-distribution is appropriate. In this case, a different value of k_C will be calculated. The values of the t-distribution that give the proper value of the coverage factor are given in a table appended to this section. Many investigators are comfortable with assuming that the coverage factor is 2.0 in all cases; however, this assumption is erroneous and sometimes leads to significant underestimates of the required coverage factor when the sample is small. Actually, it requires almost no extra effort to compute the appropriate coverage factor from the t-distribution for an actual finite sample.

Fractional uncertainty

The uncertainty defined above is essentially a dimensional number. Indeed, it must have the same dimensions as the associated measurement. In many applications, however, it is convenient or even natural to refer to the fractional uncertainty. The fractional uncertainty is the ratio of the uncertainty to the measurement, or

$$U_{\text{fract}} = \frac{U}{m} \tag{16.10}$$

Consider the following example of a typical three digit dimensional measurement with a representative one digit uncertainty,

$$316. \pm 3 \text{ m}$$

In strict scientific notation these values would be written very formally as

$$3.16 \times 10^2 \pm 3. \times 10^0 \text{ m}$$

This dimensional uncertainty in the example corresponds to a fractional uncertainty close to one percent, specifically $9. \times 10^{-3}$. Note that a number of the order of one percent or 1.0×10^{-2} has an order of magnitude of -2. A similar four digit dimensional measurement with a one digit uncertainty could be

$$3.162 \times 10^3 \pm 3. \times 10^0 \text{ m}$$

Its fractional uncertainty of $9. \times 10^{-4}$ is close to one-tenth of one percent and is of the order of magnitude -3. Note that the difference between the order of magnitude of the LSD in the uncertainty and the order of magnitude of the measurement (*e.g.*, $0 - 2 = -2$ or $0 - 3 = -3$) is roughly equal to the order of magnitude of the fractional uncertainty.

Combining uncertainties

Any standard text on uncertainty theory should show that uncertainties due to independent sources of error are combined by summing the squared uncertainties. This rule is based on the statistical result for combining variances. Statistics is roughly the application of mathematics to probability, while measurement theory and uncertainty analysis are the application of mathematics to measurement; however, in this case, the statistical result is directly transferable. The basic statistical rule resulting from considering multiple sources of random errors is

$$\sigma^2 = \sigma_1^2 + \sigma_2^2 + \cdots \tag{16.11}$$

Note in particular that the standard deviations are not combined, so

$$\sigma \neq \sigma_1 + \sigma_2 + \ldots.$$

Experimental uncertainties are analogous to statistical SDs or SSDs; therefore, uncertainties caused by independent sources of error can be combined by summing the squares, or

$$u^2 = u_1^2 + u_2^2 + \cdots + u_m^2 + \cdots \tag{16.12}$$

In words, the overall squared uncertainty equals the sum of the squared contributing uncertainties. Note that the squared uncertainty is analogous to the statistical variance. The expanded uncertainties can also be added since the coverage factors for each uncertainty should be the same. This rule has two direct applications in uncertainty theory, calculating the combined uncertainty and estimating the range of possible bias by error propagation analysis.

Combined uncertainty

Usually, the uncertainty is known to have significant contributions from both random and bias effects. The two entirely different types of uncertainty can clearly be considered to be the results of independent sources of error. Consequently, the rule for combining uncertainties applies directly, and the squared combined uncertainty sum of the squares of the two contributing uncertainties, or

$$U_C = \sqrt{U_A^2 + U_B^2} \tag{16.13}$$

In many preliminary coursework applications, only the Uncertainty A will be known, and time will not be devoted to estimating the Uncertainty B. Therefore, it will not be possible to compute the combined uncertainty, and only the Uncertainty A will be known. Indeed, for some tests, only the uncertainty A is required. In later course work and in practice, some estimate of the Uncertainty B is needed, and then the Combined Uncertainty should be computed. Indeed, some tests are only meaningful if the Combined Uncertainty is used. In the following examples, the symbol U will be used in general for an Expanded Uncertainty of any type. If the identification of the type is important, a subscript should and will be used.

Uncertainties in direct and indirect measurement

Usually the uncertainty is not somehow known *a priori* as assumed in the trivial examples presented above to illustrate significant digits. Instead, it must be calculated or estimated. In the typical undergraduate lab curriculum, there will be several cases when the uncertainty can be calculated and other cases when it must be estimated. Some cases relate to direct measurements when the data is read from an instrument or system that responds directly to the quantity being measured. An example is the direct measurement of temperature with a digital thermocouple reader and a Type K thermocouple. Other cases relate to indirect measurements when a value of y is calculated from more direct measurements (*e.g.*, x_1, x_2, etc.) by some measurement formula

$$y = y(x_1, x_2, \cdots) \tag{16.14}$$

In some cases, the uncertainty in an indirect measurement can be directly inferred from the uncertainty in the more direct measurement or measurements without much analysis. Otherwise, error propagation analysis, which is discussed in the next section, is necessary to evaluate the uncertainty in the indirect measurement.

Error propagation analysis

Both types of indirect measurements must be considered, and error propagation analysis (EPA) is the technique to determine how uncertainties in multiple contributing direct measurements affect the uncertainty in the indirect calculated measurement. The formula for EPA calculations derives directly from the formula for combining multiple sources of error. It is first recognized that a small error or deviation from the true value of an indirect measurement can be caused by multiple independent deviations of the individual more direct measurements. Specifically, a deviation in the indirect measurement can be caused by a deviation in the mth

direct measurement. This deviation can be represented by a truncated Taylor series expansion, so

$$\delta y_m = y - y_0 = \frac{\partial y}{\partial x_m}(x_m - x_{m,0}) \text{ or } \frac{\partial y}{\partial x_m}\delta x_m \qquad (16.15)$$

where y_0 and x_0 are the true values of the indirect and the mth direct measurement respectively. The partial derivatives in the error propagation formula are sometimes called the influence coefficients. Since the variance is just the average squared deviation, it is easy to compute the variance in y due to the variance in x_m. These variances can be interpreted experimentally as squared standard uncertainties and combined by the previously stated rule. The result, which is given in any adequate reference on uncertainty, is

$$u_y^2 = \left(\frac{\partial y}{\partial x_1}u_{x1}\right)^2 + \left(\frac{\partial y}{\partial x_2}u_{x2}\right)^2 + \cdots + \left(\frac{\partial y}{\partial x_m}u_{xm}\right)^2 + \cdots \qquad (16.16)$$

This formula can be applied to calculating the Uncertainty A in an indirect measurement due to Uncertainty A in the contributing direct measurements. It can also be used for estimating the range of possible bias or Uncertainty B in the indirect measurement due to the type B uncertainties in the direct measurements. Uncertainties for the two types of measurements are addressed in the next two major sections.

UNCERTAINTY IN DIRECT MEASUREMENTS
Introduction
Most experimental cases relate to direct measurements, and all cases begin with direct measurements. In some cases, the direct measurements themselves will be used in design or research, and in other cases the direct measurements will be used to calculate the needed indirect variables. The uncertainty in indirect measurements will depend on the uncertainties in the contributing direct measurements, so direct measurement should be considered first. Most of the important practical cases will be addressed and explained individually in this section. Note that direct measurements are inherently limited to single point observations. Furthermore most of the more complicated cases involve Uncertainty A rather than Uncertainty B, but both cases will be considered here beginning with the examples of Uncertainty A. For quick reference, the recommended practice for the usual direct measurements is summarized in Tables 16.3 and 16.4.

When repeated measurements are made, the Uncertainty A can be readily calculated using statistical methods. Expounding on the full range of statistical principles and techniques is far beyond the range of this text; however, the most important simple cases can be succinctly stated. In the typical undergraduate lab curriculum, both single point and multiple point experiments will be encountered. In a single point experiment the measurements are taken when all of the independent controllable variables are kept as nearly constant as possible while the fluctuating values of the dependent variable are measured. In contrast, a multiple point experiment involves deliberately changing one or more independent variable(s) and measuring the resulting values of the dependent variable.

In the case of a single point experiment, it is conventional to assume that the controlled values of the independent variables are fixed so that the fluctuations in the directly measured dependent variable, x, are essentially random. If multiple measurements are practical in this case, the Standard Uncertainty A for the measured data is calculated using the familiar formula for the Sample Standard Deviation (SSD). In other cases, multiple measurements may not be practical or convenient, and the Uncertainty A must be estimated. Procedures for calculating or estimating the Uncertainty A are presented by examples in the following minor subsection. Multiple point experiments typically result in regression models, the parameters of which are inherently indirect measurements that are discussed below.

Uncertainty A in Single Point Direct Measurements

As background for a comprehensive set of examples, consider a liquid flow that can be measured simultaneously with a digital instrument and an analog instrument. Assume the digital instrument is a vibrating U-tube flowmeter with a 5 digit display graduated in units of kg/min with 3 decimal places to right of the decimal point (*i.e.*, 00.000 kg/min). Assume also that the analog instrument is a variable area flowmeter or rotameter with finest graduations in units of 1.0 kg/min. The finest possible resolution of the digital meter would be .001 kg/min if the smallest digit were stable enough to read, and the finest possible resolution of the analog instrument would be around .1 kg/min by interpolation. Further assume that the digital instrument has both the numerical display and a connection to a computer so that repeated measurements can quickly be made by a computer-controlled data acquisition system (DAS). For convenience, assume that the instruments have been cross calibrated (*e.g.*, by the weighing tank method) and that the analog scale of the rotameter and the digital display have been adjusted to agree with the calibration.

Repeated high resolution digital measurement.

Now consider an example of determining the measurement value and uncertainty from repeated measurements. Assume that the flow is fluctuating somewhat and that digital flowmeter data in the following table were collected by the DAS and processed by the computer. Specifically, assume the 10 data in the example database were collected and processed, as shown in Table 16.2.

Obviously, the best estimate of the mass flow rate is the calculated average of 12.308 kg/min. The best estimate of the Standard Uncertainty A is the SSD, which as indicated is 0.319 kg/min. For a small sample of data, the coverage factor k_C in Equation 16.2 is computed from the t-distribution using the auxiliary parameter called the degrees of freedom,

$$DF = N - N_p \tag{16.17}$$

where N is again the number of data, and N_p is the number of parameters calculated using the experimental data. In the case of an experimental sample, the mean of the data must be estimated using the experimental average, so $N_p = 1$. In the current case of 10 data and 1 param-

Table 16.2. Example Sample of Data from Digital Flowmeter

data index	example sample of data	
1	12.062	
2	12.215	
3	12.755	
4	12.608	
5	12.061	
6	11.881	
7	11.936	
8	12.461	
9	12.430	
10	12.676	
	12.308	= average
	0.319	= calculated SSD
	0.710	= Expanded Uncertainty of data
	0.225	= estimated U_A of the average

eter, the *DF* is 9, and 2.23 is the corresponding coverage factor. Consequently, the Expanded Uncertainty A is 0.71 kg/min, and the measurement can be most thoroughly expressed as

$$12.31 \pm 0.71 \text{ kg/min}$$

Note that the uncertainty is rounded and truncated to the recommended two digits and that the LSD in the measurement is in the same place as the LSD in the uncertainty.

This example represents the case example of any repeated high-resolution measurement. Note that, in a strict sense, this is an indirect measurement since the average is calculated; however, most investigators would not quibble about this point. Obviously, this approach would apply to measurements with an analog instrument if the fluctuations are large enough to observe but slow enough to measure. In the case of analog measurements, the observer also has the sometimes difficult task of ensuring that the data is sampled randomly.

Fluctuating measurements

Next we will address the very common instances when the fluctuating measurement cannot be conveniently and/or randomly repeated. First, consider the case of reading the digital display visually. Even if the data changed only slowly, it would be difficult to make repeated unbiased readings with much reliability or resolution. Instead, it is probably better to observe the range of variation in the reasonably legible digits. If the data rate were relatively slow, it should be possible to observe the data to trend between maybe 11.9 to 12.7 kg/min. The observer should try to estimate not the absolute 100% range but the 95% main body of the range. This distinction is obviously difficult to make, but the observer should try to estimate the main body of the variation and not react to the ultimate extremes, which could be unrepresentative glitches in the data. Based on such an observation the measurement would be taken to be the observed mid-range value, and the Uncertainty A would be taken to be half the 95% range. In this example, the measurement and uncertainty would be expressed as

$$12.3 \pm 0.8 \text{ kg/min}$$

Note here that the uncertainty is limited to one digit by the effective resolution of the instrument. Also note that this result is in good agreement with the previous case. This is an example of a slowly fluctuating digital display when the range of fluctuation can be estimated.

Next, address the case when the display is changing so fast that the variation in the minor digits is an invisible flicker. In this case, the smallest reasonably stable digit would probably be an occasionally flickering 2 in the ones place. It would be almost impossible to estimate the range of the variation in the tenths place, so the variation could be as much as one unit of the smallest legible digit. The best measurement and uncertainty in this case would be

$$12. \pm 1. \text{ kg/min}$$

This measurement is actually in rather good agreement with the previous lower uncertainty cases. Here the data is assumed to range from 11 to 13 kg/min, which is only marginally different from the ranges identified before. Note again that the effective resolution limits the uncertainty to one digit. This is an example of a rapidly fluctuating digital display when the range of fluctuation cannot be estimated but the smallest reasonably stable digit is legible.

Next, consider the analog instrument. In this case the float in the rotameter probably would be observed to fluctuate in the range of 11.9 to 12.7 kg/min with some occasional excursions above or below this range. The 95% main body of the experimental variation is confidently estimated as the 11.9 to 12.7 kg/min range. The LSD is at the limit of resolution by interpolation between markings 1 kg/min apart. The measurement would again be

$$12.3 \pm 0.8 \text{ kg/min}$$

This result is of course in agreement with the hypothetical slowly fluctuating digital instrument. This is an example of a fluctuating analog display. One must assume or determine from other information that the analog instrument can respond quickly enough that the main body of the fluctuation is apparent.

Next, consider a change in the experimental conditions so that the flow hardly fluctuates. The repeated digital DAS data needs no substantial reconsideration. The data could still be collected, and the average and the now smaller SSD could still be computed. Let's assume the new result is

$$12.532 \pm 0.023 \text{ kg/min}$$

The other cases, in which repeated measurements are impractical or unreliable, do need reconsideration.

The simplest case is the nearly stable digital display. The same obvious rules apply. Estimate the range of variation if possible and report the average and the half width of the range. Presumably the result in this case would now be

$$12.53 \pm 0.02 \text{ kg/min}$$

Here, it is assumed that only the fluctuation in the one thousandths place is too rapid to follow. If the fluctuation in the hundredths place is too rapid to follow, use the smallest legible digit as the LSD in the measurement, and assume that the range of variation is one unit of this digit, so the reasonable result would be

$$12.5 \pm 0.1 \text{ kg/min}$$

The fluctuation could not be much more than 0.1 kg/min, or the digit in the tenths place would flicker too much to be legible. In such a case, the uncertainty could be almost as much as one unit in the smallest stable digit, and the least upper bound is one unit in this place. So again, the only available estimate of the uncertainty is the least upper bound, so it is used as the best available estimate of the uncertainty.

The last common example is the nearly stable analog instrument. This measurement requires some sober consideration. In the present case, the float of the rotameter would be observed to hold a steady value of 12.5 kg/min. One could incautiously presume to make a series of repeated identical measurements and calculate the SSD to have the ridiculous value of zero, or perhaps a rare visible fluctuation in the tenths place would lead to a calculated uncertainty of say .02 kg/min. The implied measurement would then be the apparently rational value of

$$12.50 \pm 0.02 \text{ kg/min}$$

However, this value is actually unreasonable because the limit of resolution was already determined to be .1 kg/min, and the rotameter could not be read to lesser uncertainty. In reality, the limited resolution has already made any actual fluctuation invisible. Some physical change in the measurement system, such as a high-resolution magnetic read-out, would be necessary to make the higher resolution measurements possible. With the original system, fluctuations at the .01 kg/min level would be invisible, so a reasonable value for the SSD cannot be calculated. Instead observe that the instrument could fluctuate as much as the level of resolution without, by definition, being noticed. In this case, the limit of resolution is .1 kg/min. This value is the least upper bound on the uncertainty and the only reasonable estimate of the uncertainty. Consequently, report

$$12.5 \pm 0.1 \ \text{kg/min}$$

The actual fluctuation could be less than 0.1 kg/min, but such a small fluctuation could not be observed with this instrument due to the limited resolution. Note that the limit of resolution could reasonably range from 1/10 to 1/2 units of the finest usable graduations. However, the resolution must surely be no more than one half the finest divisions if the divisions can actually be distinguished. This case of a stable analog instrument is the last case of Uncertainty A commonly encountered in the undergraduate lab.

Uncertainty B in Single Point Direct Measurements

Uncertainty B in direct measurements is superficially simple. Instruments cannot be perfect, and even if calibrated, the calibrations cannot be perfect. Sources of Uncertainty B include systematic bias in the output or graduation of the instrument, the uncertainty of the calibration, and operational errors that lead to possible bias. Since any uncertainty in the calibration effects every measurement, it cannot be evaluated by statistical evaluation of repeated measurements. Consequently, the uncertainty of calibration must be considered as Uncertainty B. Uncalibrated instruments will require special attention. Usually some generic estimate of the possible bias is available or the possible bias can be inferred from the resolution of the instrument. When estimating Uncertainty B, take care to distinguish Uncertainty A from Uncertainty B to avoid double counting the uncertainty. The important common cases are reviewed in the following paragraphs.

Uncertainty B in a calibrated instrument

This is the simplest case. A useful calibration process must include a report of the possible range of bias or the Uncertainty B. A calibration function or rule for correcting to implement the calibration must also be included. For example a common calibration function for a pressure gage is

$$P_{\text{CORR}} = P_{\text{OS}} + S\,P_{\text{RAW}} \pm U_{\text{CAL}} \tag{16.18}$$

where P_{OS} is the offset, and S is the scale factor. Note that the corrected pressure is now a calculated quantity and inherently an indirect measurement. The rules for indirect measurements still apply, but most investigators would not recognize this simple correction as an indirect measurement. The calibration uncertainty can be dimensional or fractional or sometimes a combination. For example the calibration of a thermocouple may be $\pm 0.2°C$ or .5%, whichever is larger.

Sometimes the instrument can be adjusted to agree with the calibration. For examples, calibration parameters can be entered into the software of a digital instrument, or an analog electronic instrument can be adjusted to reflect the scale and offset of the calibration. Occasionally, it may be possible to modify the graduation or marking of a simple analog instrument, but this situation must be unusual. If the instrument cannot be adjusted, the calibration function must be used to adjust the measurement. The uncertainty of the calibration, which is now essentially built into the instrument, is reported as the Uncertainty B.

Uncertainty B in an uncalibrated instrument

Rather often, it is not possible or necessary to calibrate an instrument. For example, a standard commercial thermocouple may be adequate for a routine measurement. In this case, it should be sufficient to rely on generic information about the range of possible bias in similar instruments. For example, one manufacturer may publish an estimated uncertainty of $\pm 2.0°C$ or 1.0% for uncalibrated Type T thermocouples. In an undemanding application, the uncalibrated thermocouples can be used with this Uncertainty B.

Also rather often, the likely Uncertainty B of a simple uncalibrated instrument can be interpreted from its resolution. For a classical example, a good quality micrometer may be graduated to .001 inch, implying that the instrument may be read by interpolation to .0001 inch. A reasonable estimate of the possible bias is then ± 0.0001 inch inferring that the instrument is graduated to comply with its possible bias. Alternatively, some manufacturers limit their .0001 inch grade calipers to instruments equipped with a vernier scale to assist interpolation. In this case, the Uncertainty B of the lower quality instrument could be as much as ± 0.001 inch. While no formal rule is available, experience shows that it is safe to assume that the range of possible bias of a simple instrument such as a caliper, ruled scale, or glass thermometer is no more than the limit of resolution. Consequently, this upper bound can be used as the Uncertainty B in non-critical applications.

Other practical sources of Uncertainty B

Finally, improper experimental procedure can introduce bias. A classic example is using a digital thermometer to measure temperature while the system is being heated. If the temperature is always increasing during the experiment, the finite resolution of the instrument will virtually guarantee that the temperature is biased too high. A better experimental plan would be to both increase and decrease the temperature. This procedure will ensure that the finite resolution results in some measurements that are too high and some that are too low. The result would be some random variation that can be addressed as Uncertainty A and not Uncertainty B. If the imperfection in the procedure cannot be corrected, the investigator is probably forced to include an additional estimated uncertainty in the Uncertainty B based on the resolution or other imperfection in technique. This correction would be necessary even if the instrument had been calibrated since the uncertainty induced by the suboptimal procedure is independent of the calibration.

Summary of Uncertainty in Direct Measurements

Table 16.3 summarizes the rules for evaluating the Uncertainty A of direct measurements in single point experiments. Table 16.4 summarizes the rules for estimating the Uncertainty B of direct measurements in single point measurements. Note that while multiple point experiments include single point direct measurements, there are no real multiple point direct measurements since all multiple point measurements are calculations such as regression models and are inherently indirect measurements.

Experimental work rarely ends with direct measurements. When directly measured data are processed, indirect measurements are evaluated, and these quantities require special attention as detailed in the next section.

Table 16.3. Summary of Rules for Uncertainty A in Direct Measurements

Nature of the Measurement	Estimate of the Measurement	Estimate of Expanded Uncertainty A	Basis for Uncertainty Estimate
repeated high resolution measurement	calculated average	calculated $U_A = k_C$ SSD	calculated from fluctuation in the data
slowly fluctuating digital or analog display	observed mid-range of fluctuation	half the estimated 95% range of variation	observed fluctuation in the data
rapidly fluctuating digital display	the smallest legible digit	one unit of the smallest legible digit	upper bound on fluctuation in the data
nearly stable analog instrument	the indicated steady value	the limit of resolution	limited resolution from 1/10 to 1/2 finest graduations

Table 16.4. Summary of Rules for Uncertainty B in Direct Measurements

Nature of the Measurement	Estimate of the Measurement	Estimate of Expanded Uncertainty B	Basis for Uncertainty Estimate
calibrated instrument	value calculated with calibration function	reported U_{CAL}	calibration
uncalibrated instrument with experimental experience	observed value	generic value from technical literature	experimental experience
uncalibrated instrument with finite resolution	observed value	reasonable estimate on order of the resolution	resolution
measurement with procedural limitation	observed value	reasonable estimate possibly on order of resolution	procedural limitations

Uncertainty in Indirect Measurements

Almost always, the needed data is not measured directly but is calculated from other more direct measurements by some measurement formula. In the formula, some output or indirect measurement is computed from some input contributing measurements. Usually, the inputs are all direct measurements, but some can themselves be intermediate indirect measurements. A general form of the measurement formula is

$$y = y(x_1, x_2, \cdots)$$ (16.19)

The result, y, is called an indirect measurement and requires special attention with respect to uncertainty. The result is typically no more certain than its inputs, and in some cases it can be considerably less certain than some of the inputs. Several situations will occur in the typical undergraduate lab sequence and in practical engineering. The next subsection is devoted to uncertainty in single point indirect measurements. For this case, the treatment of uncertainty is identical for both types of uncertainty. Several cases will typically be encountered in education and practice, and these more common cases will be addressed below by specific examples. The recommended rules are consolidated in Table 16.7, at the end of this section.

Uncertainty in Single Point Indirect Measurements

The most common single point indirect measurements are the results of ordinary arithmetic calculations: sums, differences, products, and quotients. Another common case is the average. After the general case is presented, these simpler cases will be addressed in turn by examples. Two other indirect measurements, the exponential and the logarithmic functions, are also important and will be considered. Another case, the power-law function, which should be representative of moderately complex analytical functions, will also be considered. The final example case will be the uncertainty for a complex measurement formula that must be evaluated numerically. There are two equivalent approaches to computing the Uncertainty A of direct measurements: (1) make the calculations and then evaluate the Uncertainty A as if the measurement were direct, and (2) evaluate the Uncertainty A of the indirect measurement from the uncertainties of the direct measurements. The first approach is trivial. The second approach is very simple and is discussed in the next subsection. If the second approach is taken, then there is essentially no difference in the treatment of Uncertainty A, Uncertainty B, or the Combined Uncertainty in single point direct measurements. In every case, the uncertainty in the indirect measurement is calculated from the corresponding uncertainty in the more direct measurements. Unless these more direct uncertainties are known, the uncertainty in the indirect measurement cannot be calculated. For this reason, the investigator should routinely ascertain the uncertainties in the direct measurements during the experiment. Fortunately, these uncertainties should be known or easily inferred from the direct measurements, and the corresponding indirect uncertainties can be calculated according to the following rules.

General single point indirect measurement

The general case is the foundation for the special cases discussed below. Actually, analysis of the general case is merely the straightforward application of error propagation analysis. The uncertainty is given by direct application of the combining rule expressed in terms of expanded uncertainties as

$$U_y^2 = \left(\frac{\partial y}{\partial x_1} U_1\right)^2 + \left(\frac{\partial y}{\partial x_2} U_2\right)^2 + \cdots + \left(\frac{\partial y}{\partial x_m} U_m\right)^2 + \cdots \tag{16.20}$$

where y is the indirect measurement, and Ui is the Expanded Uncertainty of the ith more direct measurement. The term "more direct measurement" is used since an intermediate indirect measurement can be used to compute the ultimate indirect measurement. Either the Uncertainty A or B or the Combined Uncertainty, U_C, may be computed with this formula. This technique is consistent with the often-cited classical paper by Kline and McClintock (1953).

An example of a general single point uncertainty is the Uncertainty B of the shaft power from a classical dynamometer experiment. In this experiment, the measured torque and speed are used to evaluate the shaft power. The torque is measured by a force transducer attached to a moment arm in the Prony brake configuration. The measurement formula is

$$\dot{W} = \omega T = (2\pi N)T$$

For a numerical example, estimate the Uncertainty B, in Watts, in the shaft power when the rotational speed is 1020. ± 10 RPM. Assume that the torque arm radius is 307 ± 2 mm and that the force is 215 ± 5.0 N. First, compute the nominal power to be:

$$\dot{W} = (2\pi N)T = \frac{2\pi 1020 \text{ rad}}{60 \text{ sec}}(0.307\,\text{m})(215\text{N}) = 7050.3\,\text{W or } 7050\,\text{W}$$

Next, interpret the quoted limits of accuracy as conventional 95% confidence interval Expanded Uncertainties and complete the table. A table organized like the example, Table 16.5, is highly recommended for ease in calculating, reporting, and documenting estimates of Uncertainty B. Observe that footnotes to the table and source notes in the table are used to explain and document the direct uncertainties used in the EPA calculation.

Typical source footnotes are citations to documents, such articles in technical journals, textbooks, or handbooks. Other common sources for generic uncertainties are commercial literature and manufacturers manuals. Typical sources for specific uncertainties are calibration reports and private communications. Such documents should be cited in the footnote and identified by a complete listing, in approved format, in the reference section. Source footnotes can also be used for brief explanations or for references to more lengthy descriptions

Table 16.5. Estimate of Uncertainty B in a Shaft Power Measurement

Measurement	U_x [a]	Influence Coefficient, $\dfrac{\partial \dot{W}}{\partial x_i}$	$U_i^2 = \left(U_{xi} \dfrac{\partial h}{\partial x_i}\right)^2$	Basis	Source
shaft speed, N	10.0 RPM	$\dfrac{\dot{W}}{N} = \dfrac{7050\ \text{W}}{1020.\ \text{RPM}} = 6.91\ \dfrac{\text{W}}{\text{RPM}}$	4800 W^2	resolution	(1)
arm length, r	2.0 mm	$\dfrac{\dot{W}}{r} = \dfrac{7050\ \text{W}}{307\ \text{mm}} = 23.0\ \dfrac{\text{W}}{\text{mm}}$	2100 W^2	measure-ment	(2)
force, F	5.0 N	$\dfrac{\dot{W}}{F} = \dfrac{7050\ \text{W}}{215\ \text{N}} = 32.8\ \dfrac{\text{W}}{\text{N}}$	27000 W^2	calibration	(3)
		sum of $U_i^2 =$	33,900 W^2		
		Expanded Uncertainty B [b] $=$	180 W		

Sources: (1) physical inspection, (2) precise measurement, see text, (3) calibrated by manufacturer (1997)
[a] Expanded Uncertainty B in individual direct measurement [b] 95% confidence limit

and explanations in the main text. Very brief notes may fit in the table itself. Otherwise, use a specific footnote for information about a particular entry.

In the text of the report, the result of the indirect measurement should be stated with due regard to the number of significant digits present in the uncertainty. In this case, the two digit uncertainty has its LSD digit in the tens place, so refer to the result as

$$7{,}0\underline{5}0 \pm 1\underline{8}0 \ \text{W}$$

Here the underlined LSD of the value is appropriately in the same place as the underlined LSD of the uncertainty. Underlining the LSD is for instructional purposes only.

Specific examples of simple single point indirect uncertainties
Sums and Differences
The sum is the simplest special case. The measurement formula is merely

$$y_S = x_1 + x_2 \tag{16.21}$$

Since both the influence coefficients in the error propagation formula are exactly unity, the formula for the expanded uncertainty is

$$U_{SD}^2 = \left(\frac{\partial y}{\partial x_1} U_{x1}\right)^2 + \left(\frac{\partial y}{\partial x_2} U_{x2}\right)^2 = U_{x1}^2 + U_{x2}^2 \tag{16.22}$$

This formula leads to a generalized approximate rule for sums. Assume the LSDs in both uncertainties are in the same decimal place. Since the two uncertainties are of the same order of magnitude, so is their sum. For example, uncertainties of 0.1 mm and 0.2 mm lead to a combined uncertainty of 0.224 mm, which would be rounded to 0.2 mm. Also note that in this case the LSD in the indirect measurement is in the same place as the similar LSDs in the two uncertainties. Next, consider the situation when the two LSDs are in so dissimilar that they are in different decimal places, so they usually would differ in value by at least an order of magnitude. In the sum of squares, the square of the smaller uncertainty is now smaller by twice as many orders of magnitude and should be entirely negligible. Consequently, the larger uncertainty dominates the uncertainty of the sum. For example, uncertainties of 0.1 mm and 0.02 mm give a combined uncertainty of 0.102 mm, which would be rounded to 0.1 mm. So the generalized rule for a sum is that the LSD of the sum corresponds to the larger LSD in the two terms of the sum. It is easy to show that the very same rule applies to a difference, hence the descriptive subscript.

Products

A product is almost as simple a case as the sum. For a product, the measurement formula is

$$y_P = x_1 \, x_2 \tag{16.23}$$

In this case, the influence coefficient for the first factor is the second factor and inversely for the second factor; consequently, the formula for the expanded uncertainty is

$$U_P^2 = \left(\frac{\partial y_P}{\partial x_1} U_{x1} \right)^2 + \left(\frac{\partial y_P}{\partial x_2} U_{x2} \right)^2 = x_2^2 \, U_{x1}^2 + x_1^2 \, U_{x2}^2 \tag{16.24}$$

This dimensional result does not lead directly to a recognizable generalized rule. To make progress, the fractional uncertainties should be considered, and dividing through by the square of the product gives

$$\frac{U_P^2}{y_P^2} = \frac{x_2^2 \, U_{x1}^2}{x_1^2 \, x_2^2} + \frac{x_1^2 \, U_{x2}^2}{x_1^2 \, x_2^2} = \frac{U_{x1}^2}{x_1^2} + \frac{U_{x2}^2}{x_2^2} \tag{16.25}$$

In words, this formula states that the squared fractional uncertainty of the product is the sum of the squared fractional uncertainties of the factors. Now, this is the reformulation that leads to a recognizable generalized approximate rule for products. If the two fractional uncertainties differ by an order of magnitude, then their sum of squares is totally dominated by the larger fraction. Consider a typical example where the first factor has three significant digits and therefore a fractional uncertainty on the order of 1%, say .03, while the second factor has four significant digits and a fractional uncertainty on the order of .1%, say .003, specifically. The fractional uncertainty of the product is .0302, which rounds to .03 and is still on the order of 1%. So, the larger fractional uncertainty, which corresponds to the factor with the fewer number of significant digits, controls. Therefore, the generalized rule for a product is that the number of significant digits of the product is the same as the number of significant digits in the factor with the fewer significant digits.

Quotients

The generalized rule for a quotient, or the result of division, is the same as for a product. However, since the development is a bit different the derivation will be given here. For a quotient the measurement formula is, of course,

$$y_Q = \frac{x_1}{x_2} \tag{16.26}$$

The formula for the expanded uncertainty is then

$$U_Q^2 = \left(\frac{\partial y_Q}{\partial x_1} U_{x1} \right)^2 + \left(\frac{\partial y_Q}{\partial x_2} U_{x2} \right)^2 = \frac{U_{x1}^2}{x_2^2} + \frac{U_{x2}^2}{x_1^2} \tag{16.27}$$

To derive the fractional uncertainties, divide through by the square of the quotient, giving

$$\frac{U_Q^2}{y_Q^2} = \frac{x_2^2 \, U_{x1}^2}{x_1^2 \, x_2^2} + \frac{x_1^2 \, U_{x2}^2}{x_1^2 \, x_2^2} = \frac{U_{x1}^2}{x_1^2} + \frac{U_{x2}^2}{x_2^2} \tag{16. 28}$$

This is exactly the same uncertainty formula as for the product, and the analogous rule prevails: the number of significant digits of the quotient is the same as the fewer number of significant digits in either the numerator or the denominator.

Averages

Averages are extremely common in experimental engineering. In this case the measurement formula is merely

$$y_{AVE} = \frac{1}{N} \sum_{i=1, N} x_i \tag{16.29}$$

where N is the number of data in the sample. The influence coefficient for every x_i is simply

$$\frac{\partial y_{AVE}}{\partial x_i} = \frac{1}{N} \tag{16.30}$$

It is assumed that the uncertainties of the individual measurements are identical. Then, the uncertainty for the average is computed by summing the squares of the N contributing uncertainties, so

$$U_{\text{AVE}}^2 = \sum_{i=1,N} \left(\frac{\partial y_{\text{AVE}}}{\partial x_i} U_i \right)^2 = \sum_{i=1,N} \left(\frac{1}{N} U_i \right)^2 = N \left(\frac{1}{N} U_i \right)^2 \qquad (16.31)$$

Where U_i is the identical uncertainty of the individual direct measurements, and the best available estimate of its Standard Uncertainty is the corresponding Sample Standard Deviation (SSD). The uncertainty of an average is then

$$U_{\text{AVE}} = \frac{U_i}{\sqrt{N}} = \frac{k_{\text{C}} \, \text{SSD}}{\sqrt{N}} \qquad (16.32)$$

The SSD will usually be reported as having two significant digits while N is known exactly, so the uncertainty of the average would usually have two significant digits.

Exponential functions

In the exponential function, the result is expressed as a power of e, so the measurement formula for the indirect measurement y_E is

$$y_E = C_E \exp\left(\frac{x}{x_{0,E}} \right) \qquad (16.33)$$

where x is the more direct measurement and $x_{0,E}$ is some normalizing denominator or some other denominator that makes the exponent nondimensional. The normalizing denominator is included to emphasize that the exponent should be inherently nondimensional. Since only one uncertainty is involved, the formula for the expanded uncertainty of y_E is merely

$$U_E = \frac{\partial y_E}{\partial x_1} U_x = \frac{C_E}{x_{0,E}} \exp\left(\frac{x}{x_{0,E}} \right) U_x = y_E \frac{U_x}{x_{0,E}} \qquad (16.34)$$

This intermediate result is better expressed in terms of a fractional uncertainty by dividing through by the indirect measurement y_E, giving

$$\frac{U_E}{y_E} = \frac{U_x}{x_{0,E}} \qquad (16.35)$$

So, in the case of the exponential function, the fractional uncertainty in the calculated result is equal to the normalized uncertainty in the exponent used in the calculation. Note that this formulation not only recognizes, but also emphasizes, that the exponent must be nondimensional while the direct measurement itself is usually dimensional.

Logarithmic function

The logarithm, which is the inverse of the exponential function, is equally common in experimental engineering. The measurement formula is

$$y_L = C_L \ln\left(\frac{x}{x_{0,L}}\right) \tag{16.36}$$

Here the argument of the logarithm includes some normalizing or nondimensionalizing denominator necessary to make the argument non-dimensional. Again, only one uncertainty is involved, and the formula for the expanded uncertainty is

$$U_L = \frac{\partial y_L}{\partial x_1} U_x = \left(C_L \frac{x_{0,L}}{x} \frac{1}{x_{0,L}}\right) U_x = C_L \frac{U_x}{x} \tag{16.37}$$

or more simply

$$\frac{U_L}{C_L} = \frac{U_x}{x} \tag{16.38}$$

In this case, the fractional uncertainty appears naturally. So, in the case of the logarithmic function, the suitably normalized uncertainty in the logarithm, which is the indirect measurement y_L, is equal to the fractional uncertainty in the more direct measurement x. Note that the previous exponential case is the inverse of this logarithmic case.

Generalizing the exponential and logarithmic cases

Recall the approximate formula for the logarithm when ε is a small number,

$$\ln(1+\varepsilon) = \varepsilon \tag{16.39}$$

Then, for a measurement m and its uncertainty U, which must be relatively small, recognize that

$$\ln\left(\frac{m \pm U}{m}\right) = \ln\left(1 \pm \frac{U}{m}\right) = \pm\frac{U}{m} \tag{16.40}$$

Then, note that the range of the variation in the logarithm is shown explicitly by rearranging the left hand side of the previous equation, so

$$\ln\left(\frac{m \pm U}{m}\right) = \ln\left(\frac{m \pm U}{m_0}\right) - \ln\left(\frac{m}{m_0}\right) = \pm\frac{U}{m} \tag{16.41}$$

where m_0 is any suitable nondimensionalizing denominator. Since the left-hand side is the possible range in the logarithm, then the fractional uncertainty in the far right-hand side is the uncertainty in the logarithm. Note that if the exponent in the exponential formula is properly interpreted as a logarithm, then the two uncertainty formulas are clearly seen to be inverses or mathematical mirror images.

Power law function

The power law function is reasonably common especially in thermal and fluid sciences. It will also be considered here as a rough representation of the more complex functions that may be encountered throughout experimental engineering. The measurement formula is

$$y_{PL} = C_{PL}\, x^n \tag{16.42}$$

Again, only one uncertainty is involved, and the formula for the expanded uncertainty is

$$U_{PL} = \frac{\partial y_{PL}}{\partial x_1} U_x = C_{PL}\, n\, x^{n-1} U_x \tag{16.43}$$

Converting to fractional uncertainties gives

$$\frac{U_{PL}}{y_{PL}} = \frac{C_{PL}\, n\, x^{n-1} U_x}{C_{PL}\, x^n} = n\frac{U_x}{x} \tag{16.44}$$

In this case, the fractional uncertainty in the indirect measurement is proportional to the fractional uncertainty in the direct measurement, and the proportionality coefficient is the exponent in the power law formula. If the exponent is around unity, the fractional uncertainties in the direct and indirect measurements will be similar, and the indirect measurement should have about the same number of significant digits as does the direct measurement. In reality, the exponent can range rather widely. For example, in just one paper on thermal anemometry, exponents ranging from .25 to 4 were encountered. Consequently, the simple rule that the indirect measurement has the same number of significant digits as the direct measurement would not always apply. This rule could be used in a pinch, but it would be better to evaluate the influence coefficient and then use the elementary formula directly. Since the influence coefficient can always be evaluated numerically, it is never too daunting a task. An example is given in the next section.

Complex numerical function

As an example of a complex numerical function, consider the uncertainty in the heat capacity of a fluid due to uncertainty in the fluid temperature. The functional relationship, especially for a liquid, can be rather complex, but the influence function can be readily computed numerically, even if the relationship is analytical but complex, tabular, or embodied in computer software. In this case some tabular data from a standard reference for the heat capacity of water are given in Table 16.6.

Table 16.6. Specific Heat of Saturated Water

Temperature K	Specific Heat kJ/kg·K
290	4.148
295	4.181
300	4.179
305	4.178

Assume the uncertainty at 300 K is desired. An adequate value for the influence coefficient can be calculated with a first finite difference formula as follows,

$$\frac{\partial y}{\partial x} = \frac{\partial C_s}{\partial T} \approx \frac{y_2 - y_1}{\Delta x} = \frac{4.178 - 4.181}{10\,K}\,\frac{kJ}{kg \cdot K} \tag{16.45}$$

Assuming a representative uncertainty of .2 K for the average temperature, the uncertainty in the heat capacity is

$$U\big(C_s\big) = \frac{\partial C_s}{\partial T} U_T = .0003\,\frac{kJ}{kg \cdot K^2}.0.2\,K = .00006\,\frac{kJ}{kg \cdot K} \tag{16.46}$$

Obviously, the negative sign is irrelevant in an influence coefficient and has been dropped. In this case, it is comforting to find that the uncertainty in the heat capacity is probably of no practical significance. Actually, when fitted to a power law model, the small value of 0.13 is the exponent on the temperature. As shown above, this small exponent is another indication of the obvious fact that the heat capacity is relatively insensitive to the temperature. This example should illustrate how the uncertainty from a single direct measurement could be computed, even if the measurement formula is complicated. If several direct measurements are involved, combine the uncertainties by the usual more general formula.

Summary of uncertainties in single point indirect measurements
The special cases above and the general case are summarized in the Table 16.7. Note that, while statistics are indirect measurements, they are relatively complicated. Consequently, the uncertainty of statistics is discussed and analyzed in a later section.

Table 16.7. Summary of Rules for Least Significant Digits (LSD)
and Standard Uncertainty, *U,* in Indirect Single Point Measurements

Type of Calculation or Relationship	Approximate Rule for Least Significant Digit (LSD) in the Measurement	Applicable Rigorous Formula for *U*
calibration	the digit in the same place as the LSD in the Expanded Uncertainty of the calibration	the statistically evaluated *U* of calibration, typically k_C SEE[a]
addition or subtraction of a pair of numbers	the digit in the same place as the LSD in the larger Expanded Uncertainty of either term	$U^2 = U_1^2 + U_2^2$
average of N data with known SSD[b]	same place as the LSD in the calculated $U = k_C u$	$U_{AVE} = \dfrac{k_C \, SSD}{\sqrt{N}}$
multiplication or division of a pair of numbers	the result should have no more significant digits than the number in the pair with the lower number of significant digits	$\left(\dfrac{U_y}{y}\right)^2 = \left(\dfrac{U_{x1}}{x_1}\right)^2 + \left(\dfrac{U_{x2}}{x_2}\right)^2$
statistics	largest digit allowing unambiguous ranking (*e.g.,* $\alpha = 4.998 < 5.000$)	general formula applies but usually rather obscure
more complicated or critical result	none unless the LSD can be estimated from worst case range of variation in the indirect measurement	$U_y^2 = \displaystyle\sum_{i=1}^{N}\left(\dfrac{\partial y}{\partial x_i} U_{xi}\right)^2$

[a]SEE = the usual Standard Error of Estimate calculated by regression packages

[a]SSD = the usual Sample Standard Deviation calculated by spreadsheets and statistics packages

Background on Multiple Point Experiments and Regression Models
Introduction
In a multiple point experiment, the value of some independent experimental variable is deliberately varied over some significant range, and data for the dependent variable(s) are measured. Various indirect measurements are possible, but most commonly a regression model is developed. Obviously since the results are calculations, all multiple point measurements are indirect; however, both Uncertainty A and Uncertainty B must still be addressed. In typical undergraduate lab courses and in most professional practice, regression is restricted to linear models. This restriction means that the model is linear in the parameters, which are then the constant and the coefficients in a polynomial. As will be shown below, it is possible to linearize some models that are originally non-linear in the parameters, so the restriction is not too severe in practice. It may be helpful to review the background of linear regression in the next section before proceeding with the analyses of the associated uncertainties.

Synopsis of linear regression
A typical linear model is

$$y_{est} = c + bx \tag{16.47}$$

Note that the model is linear in its parameters. For this model, the variation of the data with respect to the model, which is called the residual variation (RSS) is

$$\text{RSS} = \sum_{i=1}^{n} \left(y_i - y_{est} \right)^2 = \sum_{i=1}^{n} \left(y_i - c - b x_i \right)^2 \tag{16.48}$$

The two so-called "normal equations" that express the conditions for a minimum of the RSS are

$$\frac{\partial(\text{RSS})}{\partial c} = \sum_{i=1}^{n} 2 \left(y_i - c - b x_i \right) (-1) = 0 \tag{16.49}$$

$$\frac{\partial(\text{RSS})}{\partial b} = \sum_{i=1}^{n} 2 \left(y_i - c - b x_i \right) (-x_i) = 0 \tag{16.50}$$

or in expanded form

$$n c + \left(\sum_{i=1}^{n} x_i \right) b = \left(\sum_{i=1}^{n} y_i \right) \tag{16.51}$$

$$\left(\sum_{i=1}^{n} x_i \right) c + \left(\sum_{i=1}^{n} x_i^2 \right) b = \left(\sum_{i=1}^{n} x_i y_i \right) \tag{16.52}$$

Alternatively, in matrix form

$$
\begin{bmatrix} n & \sum_{i=1}^{n} x_i \\ \sum_{i=1}^{n} x_i & \sum_{i=1}^{n} x_i^2 \end{bmatrix} \begin{bmatrix} c \\ b \end{bmatrix} = \begin{bmatrix} \sum_{i=1}^{n} y_i \\ \sum_{i=1}^{n} x_i y_i \end{bmatrix}
\tag{16.53}
$$

The explicit solution for the coefficient b can be found with Cramer's rule to be

$$
b = \frac{n \sum_{i=1}^{n} x_i y_i - \sum_{i=1}^{n} x_i \sum_{i=1}^{n} y_i}{n \sum_{i=1}^{n} x_i^2 - \sum_{i=1}^{n} x_i \sum_{i=1}^{n} x_i} = \frac{\sum_{i=1}^{n} (x_i - x_{ave}) y_i}{\sum_{i=1}^{n} x_i^2 - n x_{ave}^2}
\tag{16.54}
$$

where

$$
x_{ave} = \frac{\sum_{i=1}^{n} x_i}{n}
\tag{16.55}
$$

With b known, it is easy enough to solve Equation 16.51 for c as

$$
c = y_{ave} - b x_{ave}
\tag{16.56}
$$

where, of course

$$
y_{ave} = \frac{\sum_{i=1}^{n} y_i}{n}
\tag{16.57}
$$

The two solutions above for c and b give the parameters for a least RSS linear model. These formulas are not really necessary since spreadsheet programs and statistics packages are available to do the calculations.

Uncertainty A in Indirect Multiple Point Regression Models
Uncertainty of the data

In practice, the investigator must select the form of model judiciously so that it can represent the systematic variation in the data. Usually, this step is relatively straightforward since the trend in the data is usually rather obvious in experimental engineering. Typically, it is safe to assume that the data varies randomly with respect to an appropriate model, since the model can be reliably fitted to represent the systematic variation. The random variation can also usually be assumed to approximately follow the t-distribution. Indeed this behavior is apparently universally expected. Then the standard uncertainty for the variation in the data with respect to the model is given by the statistic that is usually called the Standard Error of Estimate (SEE). The formula for the SEE is

$$\text{SEE} = \sqrt{\frac{\sum (y - y_{\text{est}})^2}{DF}}$$

(16.58)

where *DF* represents an index called the statistical Degrees of Freedom, which is the number of data less the number of parameters. For a model where the number of parameters is n_{p} and a data base where the number of data is *n*, the degrees of freedom is $n - n_{\text{p}}$. Experimentalists need hardly be concerned with the formula for the SEE since statistics packages and spreadsheet programs calculate it as part of the routine regression analysis. The role of the SEE for scatter with respect to a regression model is analogous to the roll of the SSD with respect to variation from the mean of a sample taken during a single point experiment. In particular, one may use the SEE as the Standard Uncertainty for computing the U_A for the data and compute this Expanded Uncertainty with this variation on the usual formula,

$$U_{\text{A,D}} = k_{\text{C}} \text{SEE}$$

(16.59)

Note that the coverage factor should be computed using the *DF* of the model as defined above. For the appropriate coverage factor, 95% of the data in a large model should fall within this range from the model. For convenience in this text this Expanded Uncertainty A is referred to as the Error Limits of the Data (ELD).

An appropriate use of the ELD is to compare the data with the model. An uncertainty envelope with error bounds ± the ELD from the model can be plotted, as in Figure 16.3. Since uncertainty envelopes are not universally accepted, it may be preferable to draw error bars on the data markers that are ELD in length, as shown in Figure 16.5. Nearly all the data should lie within the uncertainty envelope. It cannot be assumed that data outside the envelope are erroneous outliners, but it is reasonable to give these suspicious data some extra scrutiny. If an excessive number of data lie outside the limits, a numerical error in the calculation of the model or an inappropriate choice in the basic formula for the model should be suspected.

Uncertainty of a linear model

EPA is also readily applied to the model itself. First be advised that the uncertainties in the constant and the coefficient can be computed and usually are computed as part of the outline output from typical regression packages. The formulas for these uncertainties are themselves derived by error propagation analyses. The development of these formulas is straightforward but complicated, so their presentation is a bit beyond the scope of this text. The formulas are given in the more complete textbooks on regression analysis, such as Draper and Smith (1998). Since the uncertainties in the constant and the coefficient are available, a quick review of the basic EPA formula might lead one to think that the uncertainty in a simple linear model can be determined by an unsophisticated application of the combining rule. If this conjecture were true, the squared uncertainty in the model would be the sum of the squared uncertainty in the constant and the squared uncertainty in the model due to the uncertainty in the coefficient, or

$$u^2_{model} = u^2_c + x^2\, u^2_b \qquad \text{erroneous!} \tag{16.60}$$

This result is intuitively unsatisfactory because this uncertainty increases monotonically with x while one would expect the uncertainty to increase uniformly from the center toward the end points of the range of x.

The conjecture in the equation above is wrong because c and b are not independent. The proper approach is to first eliminate the coefficient by centering the data. Centering is redefining x and y with respect to their average values. This process can be shown to always eliminate the constant. The centered form of the model is

$$y_{est} = y_{ave} + b\left(x - x_{ave}\right) \tag{16.61}$$

Now, the uncertainty in the model is easy to represent by applying the combining rule to the preceding relationship as

$$u^2_{model} = u^2_{y-ave} + \left(x - x_{ave}\right)^2 u^2_b \tag{16.62}$$

The SEE is used as the estimate of the uncertainty in the y data. Then the uncertainty in the average of y, which is the average of n individual y data, is

$$u_{y-ave} = \frac{\text{SEE}}{\sqrt{n}} \tag{16.63}$$

Since the uncertainty in the coefficient has been determined and is available from the regression package, the uncertainty in the model can now be written as

$$u^2_{model} = \left(\frac{\text{SEE}}{\sqrt{n}}\right)^2 + \left(x - x_{ave}\right)^2 u^2_b \tag{16.64}$$

Knowledgeable readers may be disturbed by the simplicity of the previous result. It may seem unfamiliar and suspiciously simple. The unfamiliar form results because typical texts on regression that address the uncertainty of the model substitute the rather complex numerical formula for the uncertainty of the coefficient into this equation. Even after simplification the subsequent result is rather complex. This step is avoided here since the u_b is always available directly from the typical regression software package. The complex formula is never needed in practice.

The ultimate result for this uncertainty of the model is intuitively entirely satisfactory for at least two reasons. First, because it is a minimum at the average value of the x data where the information about the actual trend in y should be the best; and second, because it increases monotonically and approximately quadratically toward either end of the range in x. This trend is just the expected and observed behavior in uncertainty of a linear model. As usual, the uncertainty in the preceding equation is the Standard Uncertainty of the model. Note that this is the uncertainty due to random error or Uncertainty A. To plot the 95% error band, the appropriate coverage factor, k_C, should be used in computing the Expanded Uncertainty A, or

$$U_{A,\ model} = k_C\ u_{model} \qquad (16.65)$$

Rigorously, the coverage factor should be computed from the t-distribution using as the Degrees of Freedom (DF) the number of data less the number of parameters in the model.

The Uncertainty A for the model can be combined with the Uncertainty B to give the Combined Uncertainty. The Uncertainty B for a model is discussed in the next section. The Combined Uncertainty gives the error limits for the model, herein called the ELM, which can be plotted as an error envelope. It does not seem reasonable to use error bars on the model. An example application of the uncertainty of a simple linear model is plotted in Figure 16.3. The figure displays the relationship in a log-log model between the logarithm of the Nusselt number as the dependent or y variable against the logarithm of the Reynolds number as the independent or x variable.

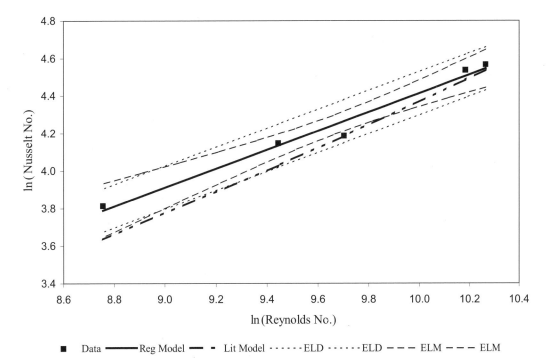

Figure 16.3. Forced Convection Data, Literature Model, and Error Envelopes
The error envelope for the data, ELD, is drawn the fixed distance U_A from the model.
The error envelope for the mode, ELM, is drawn the variable distance U_B from the model.

The regression parameters and uncertainties for the data in Figure 16.3 were computed by the Excel regression package. The curves showing the error bounds were computed in a compact block of the Excel workbook illustrated below as Attachment 1. A copy of this spreadsheet is available from the web page of one of the authors (Jeter, 2002). This example had five data points resulting in a *DF* of 3 for a linear, 2 parameter, model. For 3 degrees of freedom, the t-distribution rigorously requires a coverage factor of 3.2 as seen in Table A.2 above. Note that the error band for the data is much wider than the error band on the model, reflecting the averaging effect of the regression model.

The ELM is unfortunately not usually presented along with the regression model even if the model is proposed for design applications. This omission is unfortunate since the user would benefit from being informed about the uncertainty of the model. In experimental work the ELM is particularly helpful. Note in the previous figure that plotting the ELM makes comparing the experimental model with an alternative model from the literature trivial. From the figure it is obvious that the literature disagrees significantly with the regression model below the value 9.9 for the log of the Nusselt number. Calculating and plotting the ELM makes this comparison easy.

Uncertainties of more complex models

Uncertainties of two more complex models are considered here, a model with multiple physically distinct independent variables and a model that is a polynomial in one physical variable. Both models considered are linear in the parameters, so they are easy to formulate and evaluate using standard regression packages.

Multiple independent variables

EPA is also readily applied to the uncertainty of a more complex model with multiple independent variables that is linear in its parameters. It can be shown that centering the data by subtracting their averages from the dependent variable and the dependent variables always eliminates the constant in this case as well, so the model can always be written as

$$y_{\text{est}} = y_{\text{ave}} + b_1\left(x_1 - x_{1,\text{ave}}\right) + b_2\left(x_2 - x_{2,\text{ave}}\right) + \cdots + b_m\left(x_m - x_{m,\text{ave}}\right) \qquad (16.66)$$

As before, the uncertainty in the model is easy to formulate by applying the combining rule to the preceding relationship, so

$$u_{\text{model}}^2 = u_{y-\text{ave}}^2 + \left(x_1 - x_{1,\text{ave}}\right)^2 u_{b1}^2 + \cdots + \left(x_m - x_{m,\text{ave}}\right)^2 u_{bm}^2 \qquad (16.67)$$

The uncertainty in the average of *y*, is again computed using the SEE as the standard deviation in the formula for an average of *n* data, so

$$u_{y-\text{ave}} = \frac{\text{SEE}}{\sqrt{n}} \qquad (16.68)$$

The uncertainty in this more complicated multiple regression model is then

$$u^2_{model} = \left(\frac{SEE}{\sqrt{n}}\right)^2 + \left(x_1 - x_{1,ave}\right)^2 u^2_{b1} + \cdots + \left(x_m - x_{m,ave}\right)^2 u^2_{bm} \qquad (16.69)$$

The previous model applies to a case like modeling the heat capacity of a dense vapor that is a function of two independent variables, temperature and pressure.

Polynomial Models

The situation is a bit subtler with respect to a more complex model that is a polynomial in one variable such as this quadratic model,

$$y_{est} = y_{ave} + b_1\left(x - x_{ave}\right) + b_2\left(x^2 - x^2_{ave}\right) \qquad (16.70)$$

Surprisingly, the polynomial case is not addressed directly in the standard texts on regression analysis, so it was necessary to address this case when preparing this text (Jeter, 2003). In a polynomial model, the independent variables are linearly independent in the mathematical sense, so linear regression can find the coefficients correctly. However, because the additional variables are just higher powers of the first variable, the coefficients are strongly correlated.

Since the coefficients are correlated, the uncertainties of the coefficients cannot be considered to be independent sources of error, and the EPA formula must be modified. Specifically, the ordinary unrestricted uncertainties in the previous cases must be replaced by conditional uncertainties, and the uncertainty of the quadratic model should be written as

$$u^2_{poly\text{-}model} = \left(\frac{SEE}{\sqrt{n}}\right)^2 + \left(x - x_{ave}\right)^2 u\left(b_1|b_2\right)^2 + \left(x^2 - x^2_{ave}\right)^2 u\left(b_2|b_1\right)^2 \qquad (16.71)$$

Here $u\left(b_1|b_2\right)$ is the conditional uncertainty of b_1 given b_2, and $u\left(b_2|b_1\right)$ is the conditional uncertainty of b_2 given b_1. The conditional uncertainty of a particular coefficient is calculated after correcting the empirical data for the influence of the other coefficients. Specifically, the conditional uncertainty of the first coefficient is calculated by correcting for the influence of the second coefficient as follows,

$$y_{CORR} = y - b_2 x^2 \qquad (16.72)$$

Standard regression analysis of the corrected dependent variable data will now give the conditional uncertainty of the coefficient, b_1. The analogous formula is used to correct for the influence of b_1 before computing the conditional uncertainty of b_2. Higher order polynomial models are addressed by extending this technique to find the conditional uncertainty of each coefficient.

An example of application of the equation is a polynomial regression model such as the quadratic Clausius-Clapeyron model developed in the vapor pressure experiment or the quartic calibration model used for the thermal anemometer. Since the uncertainties of the coefficients are always computed by standard commercial regression packages, it is straightforward to calculate and plot the Expanded Uncertainty A according to the usual formula,

$$U_{A,\,model} = k_c\, u_{poly\text{-}model} \qquad (16.73)$$

The spreadsheet block illustrated in Attachment 1 is designed for applications with as many as four completely independent variables, and it can be used for a fourth degree polynomial. The results for a generic quadratic model are shown in Figure 16.4. This spreadsheet block available from the author's web page (Jeter, 2002) should make it easy to compute and plot the Uncertainty A for any regression model.

Uncertainty B for Regression Models

The Uncertainty B for a multiple point indirect model is calculated for each point by applying EPA to the more direct measurements at that point. The calculation is exactly the same as the calculation described and exemplified in the section on Uncertainty B in single point measurements. A scrupulous investigator would calculate the Uncertainty B for every point. A simplification is to calculate the representative Uncertainty B for a midrange value of the indirect measurement and apply that uncertainty to the entire data base. Once this Uncertainty B is available, it can be used to compute the Combined Uncertainty as described in the next section.

Combined Uncertainties for Regression Models

To compute the Combined Uncertainty, first compute the Expanded Uncertainty A of the model as detailed above. Then, compute the Expanded Uncertainty B of the indirect measurement by applying EPA to the overall measurement system. Then, compute the Combined Uncertainty of the model, as

$$U_{C,\,model}^2 = U_{A,\,model}^2 + U_B^2 \qquad (16.74)$$

Note that the spreadsheet block illustrated in Attachment 1 has a column for the Uncertainty B of the model. The U_B can be a representative constant value; or, preferably, it can vary as the direct measurements vary.

In Figures 16.4 and 16.5 that show example regression models, the uncertainty band for the data is delimited by dotted lines that plot the Expanded Uncertainty A of the data above and below the model. For the data, the Uncertainty A is a constant. Specifically, the Expanded Uncertainty A of the data is the constant computed by the usual formula

$$U_{A,\,data} = k_c\, \text{SEE} \qquad (16.75)$$

The uncertainty band for the model is delimited by broken lines plotted the Combined Uncertainty of the Model above and below the regression model. The Uncertainty A of the

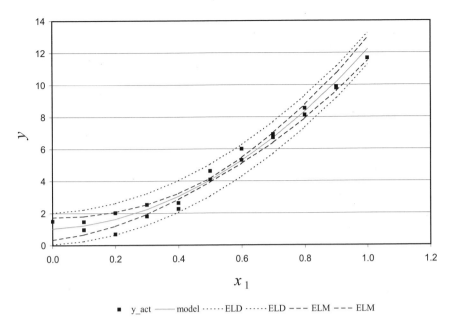

Figure 16.4. Example Plot of a Quadratic Model with the Uncertainty Band for the Data, which has the Limits ELD, and the Uncertainty Band for the Model, which has the Limits ELM

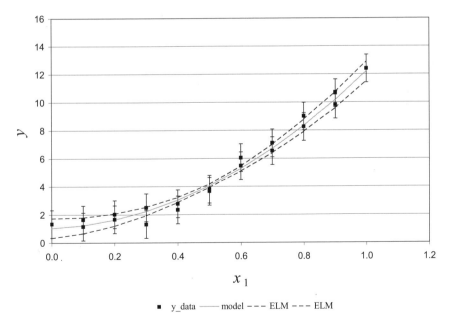

Figure 16.5. Example Plot of a Quadratic Model with Uncertainty Error Bars for the Data and an Uncertainty Envelope for the Model. The length of the error bars is the Uncertainty A or ELD of the data. The width of the uncertainty envelope is the Combined Uncertainty or ELM of the model.

237

model varies with x in a roughly quadratic fashion. The Uncertainty B of the data can be taken to be a constant; or, preferably, it should vary with x. To plot the error limits on the model, the Expanded Combined Uncertainty of the model is computed with Equation 16.74. The Expanded Uncertainty A of the model needed in Equation 16.74 is computed using Equation 16.65. The result is the Error Limit of the Model (ELM) plotted in the regression examples. The details for calculating the uncertainties of regression models are summarized in Table 16.8.

Table 16.8. Summary Guidelines for Computing the Uncertainties in Regression Models

Type of Uncertainty	Formula	Applicable Equation
Uncertainty A in the data or ELD	$U_{A,data} = k_c\,SEE$	16.75
Uncertainty B in the data	general formula for U_B by EPA	16.20
Uncertainty A in the model*	$U_{A,\,model} = k_c\,u_{model}$	16.65
Uncertainty B in the model	general formula for U_B by EPA	16.2
Uncertainty C in the model or ELM	$U^2_{C,model} = U^2_{A,model} + U^2_B$	16.74

*Use formula for the u of a polynomial model, Equation 16.71, when necessary.

Uncertainty and Significant Digits in Statistics

Some statistics present no special difficulty in identifying the least significant digit. Other statistics generated in regression analysis, such as the Standard Error of Estimate, the Coefficient of Determination, and the probability called the alpha risk, require specific consideration.

The Standard Error of Estimate is effectively the square root of the averaged squared deviation from a regression curve and is analogous to the sample standard deviation for a single point measurement. Consequently, it is a Standard Uncertainty and should typically be expressed with two digits. The second digit is included to allow application of the two-digit rule and to allow an accurate calculation of the percent error. Recall that the Standard Error of Estimate is typically interpreted as the Standard Uncertainty A of the data with respect to a regression model.

The Coefficient of Determination, also called the R-Squared, is a statistic that quantifies how well a regression model represents the experimental data. Its maximum value is unity, and in engineering applications it can be and usually is very close to unity. Two issues arise with this statistic: How many significant digits of an R-Squared should be displayed, and What difference constitutes a significant difference between two R-Squared values when they are compared. Both of these issues would be easily resolved if the uncertainty of the R-Squared were routinely calculated and used. Since the R-Squared is just another indirect measurement, computing its uncertainty is straightforward. The only complication is the obvious fact that the measurement function is somewhat complex, so most investigators would resort to numerical evaluation of the influence coefficient. This complication is minor, and the uncertainty can be readily computed with a little extra effort. Unfortunately, the uncertainty of the R-Squared is not generally recognized for use as a criterion for displaying or comparing R-Squared values.

The literature does not seem to address the issue, so it is probably not trivial to generalize the behavior of the uncertainty of the R-Squared in practical applications. The fractional uncertainty has been computed in several specific practical cases, and the value was found to vary over many orders of magnitude from the order of 1% to .001%. Consequently, no analytically based simple rule can be recommended here. Furthermore, it is probably not appropriate to estimate the uncertainty with the rigorous method because this approach is not recognized as conventional. Readers would be distracted by too much attention to this relatively minor point.

Since no simplified analytical rule is available and the rigorous evaluation is not conventionally accepted, some practical stopgap is necessary. Consider the case where it is necessary to display and to compare and rank two or more very close values of the R-Squared (*e.g.*, to compare 0.99983896 and 0.999968984). If the underlying data is not very accurate (*e.g.*, to three digits), it may seem inappropriate to report the four significant digits necessary to rank the two R-Squared values based on this relatively uncertain three digit data. Actually when the uncertainty of the R-Squared is evaluated in this case, the fractional uncertainty does actually justify six significant digits. To resolve this issue, observe that statistical theory assumes that all numbers are known to enough digits to allow all data to be ranked with no ties. The mathematical concept is that every number is a real number with an essentially infinite number of digits. A reasonable practical rule for displaying statistics results. Assume first that it is appropriate to report at least two digits so that an accurate percentage can be reported. Then assume that enough digits, but no more, can be reported to unambiguously rank several values if several statistics are being compared (*e.g.*, report 0.9998 and 0.9999).

For probabilities such as the alpha risk, report at least two digits so that an accurate percentage can be computed, and report enough digits to allow an accurate ranking if the probability must be compared with another probability or a prescribed criterion. For example, if the alpha risk is computed to be .049798, then report it as .0498 not .05 if it is important to show that it is at least marginally less than 5%. As a reasonable general rule, assume that just enough digits of a probability can be reported to make an accurate ranking with no ties.

Examples

Some examples of applying uncertainty principles to identifying the significant digits are detailed in Tables 16.9 and 16.10. Table 16.9 is a representation of either poor or no apparent attention to significant digits, and Table 16.10 is an improvement to a professional level of presentation. The data is meant to represent raw mass flow and temperature data used to determine a heat rate calculated by

$$\dot{Q} = \dot{m} C_P (T_{out} - T_{in}) \qquad (16.76)$$

Specify *a priori* that the flow meter can be read to the sixth decimal place but that it is calibrated to only ± .0003 kg/sec. Specify as well that the thermometers can be read to the hundredths decimal place but are corrected by a calibration function with uncertainty of .2°C, the tenths place. Note that the specific heat varies negligibly over the entire temperature range so it is almost completely insensitive to the much smaller uncertainty in temperature.

The specified uncertainties and the simplified rules are used to format Table 16.10. The raw mass flow data are assumed to be resolvable to the sixth place, so a consistent six-digit numerical format is used for the entire column. An argument could also be made to round the raw data to four places to agree with the corrected data. In accord with the uncertainty of the flow meter calibration, the corrected values are rounded to the fourth place. In accord with the uncertainty in the temperatures, the least significant digits are in the tenths place for both corrected temperatures. Adding the offset 273.15, which is an exact value, does not move the least significant digit. The temperature difference and average temperature are significant to the tenths place since these values are computed by mere subtraction and addition. The specific heat remains certain to four digits since it is seen to not vary with respect to the uncertainty in the temperature data. Finally, the computed heat rate, an indirect measurement, has two significant digits since the mass flow rate, with two significant digits, has the fewest significant digits of the factors in its computation. The table has also been improved by italicizing all the math symbols.

Table 16.9. Example of Inattention to Significant Digits and Numerical Format

raw	corr	raw	corr	raw	corr			Cp	heat
m-dot	m-dot	T-in	T-in	T-out	T-out	delta -T	T-avg	at T-avg	rate
kg/sec	kg/sec	C	C	C	C	C-deg	C	kJ/kg-K	kJ/sec
0.006788	0.006584	24.21	24.89	45.3	46.69	21.79	35.79	4.179	0.600
0.008593	0.008335	24.33	25.02	40.23	41.47	16.45	33.24	4.179	0.573
0.008934	0.008666	24.23	24.91	39.8	41.02	16.11	32.97	4.179	0.583
0.00948	0.009196	24.12	24.80	36.84	37.98	13.17	31.39	4.179	0.506

Table 16.10. Example of Good Practice in Significant Digits and Numerical Format

raw mass flux	corr mass flux	raw inlet T	corr inlet T	raw outlet T	corr outlet T	T diff	avg T	Cp	heat rate
$\dot{m}$ kg/sec	$\dot{m}$ kg/sec	T_{in} °C	T_{in} K	T_{out} °C	T_{out} K	ΔT K	T_{avg} °C	at T_{avg} kJ/kg·K	Q kJ/sec
.006788	.0066	24.21	298.0	45.30	319.8	21.8	35.8	4.179	0.60
.008593	.0083	24.33	298.2	40.23	314.6	16.5	33.2	4.179	0.57
.008934	.0087	24.23	298.1	39.80	314.2	16.1	33.0	4.179	0.58
.009480	.0092	24.12	298.0	36.84	311.1	13.2	31.4	4.179	0.51

APPENDIX A. The Coverage Factor, k_C, for 95% Confidence Interval
Tabulated for Various Degrees of Freedom, *DF.*

In general, the Degrees of Freedom is the number of data points less the number of parameters calculated from the data, so when there are *p* parameters and *n* data, $DF = n - p$.

When evaluating the Expanded Uncertainty of an average of a sample of *n* data, $DF = n - 1$.
When evaluating the Expanded Uncertainty of a linear model based on *n* data, $DF = n - 2$.
When evaluating the Expanded Uncertainty of a model involving
p parameters based on *n* data, $DF = n - p$.

DF	k_C	DF	k_C	DF	k_C
1	12.71	12	2.18	40	2.02
2	4.30	14	2.14	50	2.01
3	3.18	16	2.12	60	2.00
4	2.78	18	2.10	70	1.99
5	2.57	20	2.09	80	1.99
6	2.45	22	2.07	90	1.99
7	2.36	24	2.06	100	1.98
8	2.31	26	2.06	110	1.98
9	2.26	28	2.05	120	1.98
10	2.23	30	2.04	infinity	1.96

The statistics in this table were computed using
the Excel spreadsheet function TINV in this cell formula:
= TINV(.05, DF)

References

Draper, N. R. and H. Smith, *Applied Regression Analysis*, 3rd ed., 1998, John Wiley and Sons, New York.

Jeter, S. M., 2002, "Example Spreadsheet, ELM.xls," ME 4053 Engineering Systems Laboratory, the George W. Woodruff School of Mechanical Engineering, Georgia Institute of Technology, Atlanta, GA, 4 January, document is available on line at <www.me.gatech.edu/sheldon.jeter>.

Jeter, S. M., 2003, "Evaluating the Uncertainty of Polynomial Regression Models Using Excel," to be published in the 2003 ASEE Annual Conference Proceedings, June 2003.

Kline, S. A. and F. A. McClintock, 1953, "Describing Uncertainties in Single-Sample Experiments," *Mechanical Engineering*, vol. 75, pp. 3–8.

Massey, B. S., 1986, *Measures in Science and Engineering*, Halsted Press, New York.

Palmer, A. de F., 1912, *The Theory of Measurements*, McGraw-Hill, New York.

Shoemaker, D. P., 1996, *Experiments in Physical Chemistry*, McGraw-Hill, New York.

Skoog, 1969, *Fundamentals of Analytical Chemistry*, Holt, Rinehart and Winston, New York.

Taylor, B. N. and P. J. Mohr, 2002, "The NIST Reference on Constants, Units, and Uncertainty," NIST Physics Laboratory, NIST, Gaithersbery, MD, 23 July 1999, this reference is available online at <http://physics.nist.gov/cuu/Uncertainty/index.html>.

Attachment 1. A Spreadsheet Block for Calculating and Plotting the ELD and ELM

Sheet to Compute and Plot the Uncertainty of a Model, SMJ Nov 2001, updated 3 July 2002

User must insert and/or update data coded with yellow. User should format these and other cells for neatness or legibility as needed.

Instructions:
(1) Copy this sheet to your experimental *.xls workbook.   (2) Insert the experimental data into block 4.
(3) Insert the regression results into block 2.   (4) Select the desired coverage factor in Block 3.
(5) Update the cell ranges to compute the averages in Block 3.   (6) Identify by cell formula the max and min X1 values in Block 5.
(7) Insert data or formulas for other XN data in Block 5.   (8) Insert optional data for U_b in Blocks 4 and 5.
(9) Plot experimental data points with green block in Block 4.   (10) Plot model and limits with green block in Block5.

1. Summary Data:

The averaged U_a of model = 0.152   The averaged U_c of model = 0.152
The constant U_a of the data = 0.380

2. Block of Data from Regression

User must insert the following data from from the regression block.

Constant =	1.43743	
Coefficient B1 =	0.96190	0.00142 = Std error of B1
Coefficient B2 =	0.00000	0.00000 = Std error of B2
Coefficient B3 =	0.00000	0.00000 = Std error of B3
Coefficient B4 =	0.00000	0.00000 = Std error of B4
Std Error of y Est =	0.17056	
n, number of data =	12	
p, number of parameters =	2	
coverage factor, kc, by t-dist =	2.23	

3. Block of Calculations

User must specify the desired coverage factor to be used. Rigorous value is in cell D23.
coverage factor, kc, used = 2.23   <-----User must select.

User must update the cell ranges in the following four formulas to calculate the correct averages.
Average value of X1 = 44.54383
Average value of X2 = 0.00000
Average value of X3 = 0.00000
Average value of X4 = 0.00000

4. Block of Experimental Data and Results, plot the data with the block in green

User must insert the complete set of y and x data from the experimental data set into following block. Add rows as required.
User may insert column of data for Expanded Uncertainty B, uncertainty due to possible bias, if desired.

Experimental Data, insert at least one zero in every otherwise unused XN column

Y data	X1 data	X2 data	X3 data	X4 data							Lower	Upper	Lower	Upper
	X1	X2	X3	X4	Ub of model	Ua of model	Uc of model	X1 data	Y data	regress model	E-limit on model	E-limit on model	E-limit on data	E-limit on data
-8.737	-10.698	0.000	0.000	0.000	0.010	0.206	0.206	-10.698	-8.737	-8.853	-9.059	-8.647	-9.233	-8.473
0.781	-0.880				0.010	0.180	0.181	-0.880	0.781	0.591	0.410	0.772	0.211	0.971
10.382	9.302				0.010	0.156	0.156	9.302	10.382	10.385	10.229	10.542	10.005	10.765
19.988	19.309				0.010	0.136	0.136	19.309	19.988	20.011	19.875	20.147	19.631	20.391
29.583	29.308				0.010	0.120	0.120	29.308	29.583	29.629	29.509	29.749	29.249	30.009
39.266	39.434				0.010	0.111	0.111	39.434	39.266	39.369	39.258	39.480	38.989	39.749
48.948	49.531				0.010	0.111	0.111	49.531	48.948	49.081	48.970	49.193	48.701	49.461
58.663	59.688				0.010	0.120	0.120	59.688	58.663	58.852	58.731	58.972	58.471	59.232
68.340	69.723				0.010	0.135	0.136	69.723	68.340	68.504	68.368	68.640	68.124	68.884
78.093	79.810				0.010	0.156	0.157	79.810	78.093	78.207	78.050	78.363	77.827	78.587
88.056	89.923				0.010	0.180	0.181	89.923	88.056	87.935	87.754	88.115	87.555	88.315
98.048	100.076				0.010	0.207	0.207	100.076	98.048	97.701	97.494	97.908	97.321	98.081

5. Block of Uniformly Spaced Data for Plotting wrt X1, plot model and limits with the block in green

User must insert cell references to identify the maximum and minumum X1 values in the cells on the next row.
max X1 = 100.076   min X1 = -10.698   1.11E+01 = computed delta X1
Spreadsheet will compute uniformly spaced X1 values. User must code columns for corresponding values of X2, X3, and X4.
User may insert column of data for Expanded Uncertainty B, uncertainty due to possible bias, if desired.

X1	X2	X3	X4	Ub of model	Ua of model	Uc of model	X1 data	regress model	Lower E-limit on model	Upper E-limit on model	Lower E-limit on data	Upper E-limit on data
-1.1E+01	114.45			0.010	0.206	0.206	-1.1E+01	-8.853	-9.059	-8.647	-9.233	-8.473
3.79E-01	0.14			0.010	0.177	0.178	3.8E-01	1.802	1.625	1.980	1.422	2.182
1.15E+01	131.26			0.010	0.151	0.152	1.1E+01	12.458	12.306	12.609	12.078	12.838
2.25E+01	507.79			0.010	0.130	0.130	2.3E+01	23.113	22.983	23.243	22.733	23.493
3.36E+01	1129.74			0.010	0.115	0.115	3.4E+01	33.769	33.653	33.884	33.389	34.149
4.47E+01	1997.11			0.010	0.110	0.110	4.5E+01	44.424	44.314	44.534	44.044	44.804
5.58E+01	3109.89			0.010	0.115	0.116	5.6E+01	55.079	54.964	55.195	54.699	55.459
6.68E+01	4468.09			0.010	0.130	0.131	6.7E+01	65.735	65.604	65.865	65.355	66.115

Part Three:
Writing on the Job

Jeffrey Donnell
and
Colin MacDougall

CHAPTER 3.1

GUIDE TO REPORTS IN THE WORKPLACE

Introduction

Reports that are prepared for workplace projects superficially resemble the reports that students prepare for undergraduate projects. That is, they use the same kinds of section headings, they present information in roughly the same order, and they use the same kinds of plots and diagrams to display results. Workplace reports differ from student reports because they are more complex, and this complexity can involve both the subject matter and the audience.

Subject Matter

Undergraduate laboratory projects almost always present students with well-defined goals and procedures, whereas workplace projects do not always do this. With the exception of form-driven quality tests, workplace projects will likely require you both to determine your project's technical goals and to define how particular tasks help you meet those goals. You will usually present these definitions in your report's Introduction section, which consequently will tend to be longer than the Introduction sections of your undergraduate reports.

Audience

Undergraduate reports are written for your instructors. These are people who you know and who understand your projects even before you submit your reports. Workplace reports, in contrast, may be read by people other than your supervisor. You may not know these people, and you cannot assume they are experts in your field. Yet you should make sure your reports are accessible to such outside readers because they can have a significant impact on your work—they may include government officials who must approve your work, corporate lawyers who might defend your work, or finance officials who control your project's funding.

You can best solve the problems of audience and subject matter by writing reports that are complete and by using images that provide context. Completeness in workplace reports is managed at the beginning of the document. Here your Introduction should summarize the task you received from your client or supervisor, and it should explain how your project steps help you to meet the objectives that have been assigned to you.

Presenting Images

Many of your readers will be unfamiliar with your project. They will have trouble visualizing both the things you have made and the environment where these things will be used. You can help these readers understand your work by displaying images, such as photographs, drawings, and diagrams. As you design images for the different sections of your reports, you should assume that your figures do different jobs for the reader and provide different kinds of information at different points in your report. In an Introduction, for example, your job is to name your topic and orient your reader, so figures should display with few details the subject of your work or the object you were asked to modify. A Result or Design section should provide substantial information about your work, so figures should display, for example, detailed views of the device you have designed or examined. In the same way, an Analysis section might describe safety calculations, so it might display a mathematical simplification of the system you have developed or tested.

To assure that non-expert readers can understand your work, it is useful to provide context information for all sections of your reports. This often requires that you use pairs of illustrations, where a diagram of new, technical information is displayed beside a photograph or drawing of the device or system as it appears in common use. The use of paired illustrations greatly simplifies the job of the non-expert reader in a technical report. You can further simplify that reader's task by preparing figure descriptions that define what is displayed in your paired figures.

In addition to selecting figures that present an appropriate amount of information, it is important to label the important components in those figures and describe them in text. Those text descriptions should be brief, and they should explicitly call out and explain the labeled parts of the drawings or photographs.

Below we present three sets of paired figures from a sample report that will be presented in a later chapter. Using these figures and their text descriptions, we will explain how you can easily and clearly present complex work to readers who are not professionals in your field.

An introductory figure

The first job of a report's Introduction section is to name the subject of study and define the problem that the project addresses. Because readers are not always technical experts, that definition needs to be concrete, and it ought to connect the technical task to an object or concept that most readers should have seen or heard of before. Figure 1 shows the opening page of our example report on the redesign of a modular bridge expansion joint. Here, four sentences and one figure work together to present all of this information.

1a) Name the subject for non-experts.
The first sentence defines Modular Bridge Expansion Joints in functional terms, and it cites Figure 1, which shows on the left a typical expansion joint on a highway bridge.

1b) Identify the subject for experts.
The second sentence defines the expansion joint components that are important in this project. These components are labeled in the diagram on the right side of the figure, and these labels correspond to the terms used in the text. That diagram is aligned with the photograph in such a way that the roadbed, joint, and concrete barriers have roughly the same arrangement

1A

1B

1C

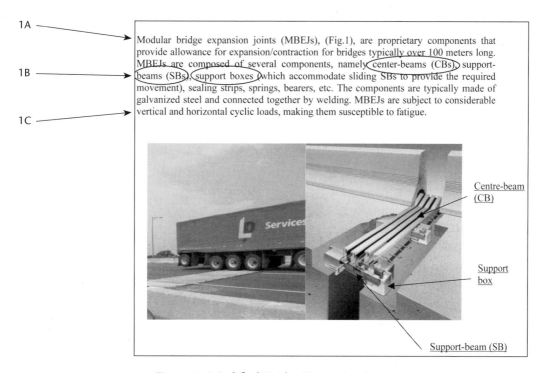

Modular bridge expansion joints (MBEJs), (Fig.1), are proprietary components that provide allowance for expansion/contraction for bridges typically over 100 meters long. MBEJs are composed of several components, namely center-beams (CBs), support-beams (SBs), support boxes (which accommodate sliding SBs to provide the required movement), sealing strips, springs, bearers, etc. The components are typically made of galvanized steel and connected together by welding. MBEJs are subject to considerable vertical and horizontal cyclic loads, making them susceptible to fatigue.

Centre-beam (CB)

Support box

Support-beam (SB)

Figure 1. Modified Bridge Expansion Joint

in each image. This alignment makes it easy for all readers to see how the diagram maps onto the photograph of a typical joint.

1c) Define the problem.

The third and fourth sentences define the problem, with sentence three describing the materials and fastening methods and sentence four naming the specific problem of fatigue under cyclic loading, as represented by the trailer shown in the right-side photograph.

A design or results figure

Results, like introductory information, are best presented using a combination of diagrams and words. Figure 2, the design presented by our sample report, offers an excellent example of such a presentation, as it is complete, direct, and short. The diagram is lightly dimensioned, as is reasonable for an initial discussion of a design concept. The diagram is fully labeled; the four components of the proposed design are labeled and called out in the text description. That description fills three sentences. The first cites the figure and defines its topic. The next two sentences list and explain the labeled components of the diagram; the circled terms call out labels in the drawing. Because an existing device is being modified, the explanations focus on methods and points of attachment. Specifically, the discussion of the FRP box section states that it is to be "epoxied to the bottom flange of the steel beam."

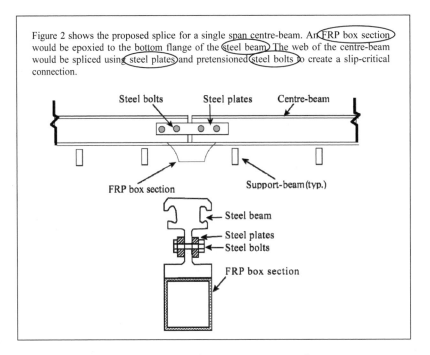

Figure 2 shows the proposed splice for a single span centre-beam. An FRP box section would be epoxied to the bottom flange of the steel beam. The web of the centre-beam would be spliced using steel plates and pretensioned steel bolts to create a slip-critical connection.

Figure 2. Figure and text presenting a design

A background figure

A review of the prior art with regard to a problem can help to clarify a problem and to explain the advantages of the solution that your team might propose. In Figure 3, the prior art is presented visually, with beam splices represented both photographically and diagrammatically. The figure description here is very short, but the photograph and the diagram have been prepared such that the splice and the bolts very prominently display the points of alignment, while the text identifies fatigue as a problem in this form of splice.

Tables

To make tabular information accessible for the non-expert, it is best to keep tables tightly focused and to describe them thoroughly. Table 1 in Figure 4 presents the results of a simple analysis, and it demonstrates how to do both of these things. In this case, a calculation was performed to determine the length of a splice that would be attached to reinforce a support beam, and the results were presented in tabular form. The table contains two kinds of information: the deflection of an intact beam and the calculated deflection of a spliced beam. The goal would be to determine the size of a splice so that the beam-splice system deflects no more than an intact, unmodified beam. The table compares the deflection of an unmodified, intact beam with the deflections attributed to beams having three different sizes of splice. This table justifies a decision to use a splice length of 400 mm because this yields a deflection close to but less than the deflection of an unspliced beam. This table is small, and it is efficient. It presents an analytical

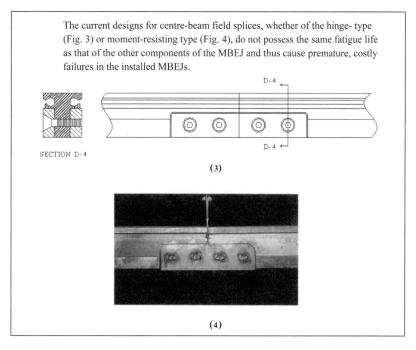

The current designs for centre-beam field splices, whether of the hinge- type (Fig. 3) or moment-resisting type (Fig. 4), do not possess the same fatigue life as that of the other components of the MBEJ and thus cause premature, costly failures in the installed MBEJs.

D- 4

SECTION D- 4

D- 4

(3)

(4)

Figure 3. A Background figure aligns a diagram with a photograph

The model was first used to determine the length of FRP required to ensure the stiffness of the spliced beam is similar to that of an intact beam. In this case, a GFRP beam with $E_{11} = 120$ GPa was assumed. Table 1 shows the results.

Table 1 – The effect of FRP section length on the maximum deflection of the spliced beam.

Beam Condition	Length of FRP section (mm)	Maximum Deflection (mm)
Intact	-	1.515
Spliced	200	1.982
Spliced	300	1.741
Spliced	400	1.484

The maximum deflection of the intact beam under the truck loading is 1.515 mm. The results indicate that the GFRP box beam (E = 120 GPa) must be at least 400 mm long to ensure that the maximum deflection of the spliced beam is similar to that of an intact beam. The typical span of a centre-beam is 1 metre, so a 400 mm FRP beam would still have 300 mm clearance on each side for installation in the field.

Figure 4. An example results table with text description

result that justifies a decision to set the splice size at 400 mm. It does this by offering two kinds of comparative information—the benchmark value of 1.515 mm deflection for an unmodified, intact beam, and the comparison deflections for splices of 200 and 300 mm.

The text description of the table is complete, but the description is brief because each sentence performs a specific task. The first presents the goal of the calculation, and the second presents the assumptions that were used. The three sentences of the second paragraph, following the table, speak directly to the contents of the table, calling out the benchmark value of the intact beam, the assumed value of E11, and the selection of 400 mm as the best size for the splice. The analytical result is captured in a table of five rows, while the text fully speaks to the table in the space of seven lines.

Chapter 3.2

Guide to Report Organization

Workplace reports are organized and assembled according to the same format scheme as is used in undergraduate student reports. Workplace reports differ from undergraduate reports due to matters of scale rather than of kind. Accordingly, the sample report presented here is subdivided into roughly the same headed sections as are found in student reports:

Introduction
Methods and Organization
Results
Analysis
Conclusion

Below we describe briefly what information you should present in each of the main headed sections of a typical report. Following this we present a sample project report that we have annotated to call attention to the way information is presented in each of these sections.

Introduction

The Introduction to a report should explain what work has been performed and why it has been performed. In most cases, report introductions are built up from five pieces of information:

1. The name of the Topic
2. The Motivation or Need for the work
3. Related Background and Problem
4. The Goal of your work
5. The Scope of the report

This information should be as brief as is reasonably possible. To present this information, it is often helpful to compile your Introduction by recording your answers to these questions:

1. What did you work on?
2. Why was it necessary to do this work?
3. Has other work been done in this area?
4. What did you attempt to accomplish?
5. How much of your work is presented here?

This list reflects the questions that readers ask most often when they begin to read technical reports, and it presents the most common way to order these statements. However, small variations are common and reasonable. For example, when your report describes a completed project, the Goal and Scope of the report will probably be the same. In the same way, the Motivation and Background statements will be the same when you address a problem that others have tried and failed to solve. In our sample design report, these questions are addressed over the course of several pages, including both the **Introduction** section and the **Requirements for Fatigue-Resistant MBEJ Splice** section.

Workplace projects are commonly undertaken at the request of a client or a supervisor. When this is the case, your Introduction should first thoroughly review the specifics of that request. After you complete that review, you should add any other information that is pertinent to the task, such as attempts that others may have made to solve the problem. When you present your own background research as well as your client's description of a problem area, you should present the client's problem description first, and you should make it clear where the client's problem description concludes and your input begins. A simple way to draw this distinction is to place your contribution in a separate paragraph, and to open that paragraph with a transitional statement such as this: "In order to address these concerns, it is important to consider the following additional factors...."

Project Formulation/Goals

Your last job in an Introduction is to identify the particular goals of your work and the particular challenges you addressed in order to meet those goals. This is commonly offered in the form of a list that describes your project work and that corresponds to the list of sections in the rest of your report. In this report, the section **Requirements for Fatigue-Resistant MBEJ Splice** presents a concise formulation of the project's goals.

Workplace reports do not always provide separate **Methods** sections because workplace projects do not always involve experimentation. In design and analysis reports, such as that presented here, methods information is distributed through the task and analysis sections of the document, while only a general overview of your project's organization is presented at the point of the task formulation. In this example report, that organizational information is compressed to two sentences at the end of the section **Requirements for Fatigue-Resistant MBEJ Splice**.

Design Description or Results Description

Because projects can take many forms, the findings or results presentations can have many flavors. On some projects, you will present objects or photographs of facilities that you have examined, while for other projects you might use plots and data tables to display your accomplishments. In either case, your results are usually best presented visually; your task in the report is to display those results effectively and to explain them so that readers can see what is significant about them.

The standards for preparing displays are described elsewhere in this book. Here it is sufficient to remind you that figures need to be sized so that they are large enough to be clearly visible in printed form. It is also a good idea to highlight important areas on your images by imposing one or two arrows or circles. When you do this, you need to explicitly speak to those highlights in the text description of your display.

Any time you display a figure, you should provide a descriptive caption, which answers the question "What is this?" For professional reports you should also provide a text description of the display that explicitly defines what is in the display and what readers are expected to derive from it. To do this efficiently, you should describe figures according to this simple checklist.

1. Cite the figure by number in your text.
2. State the figure's purpose.
3. List the labeled components of the figure.
4. Discuss the figure:
   a. Explain what the image demonstrates (for plots or tables).

   or

   b. Explain how the device operates (for devices or photographs of devices).

Other types of discussion are acceptable, of course, depending on the image.

Our sample report offers an excellent design presentation in the headed section **Proposed FRP Splice Design**.

Analysis or Validation

Results presentations usually are supported by some type of discussion or analysis that either explains the results that have been obtained or that validates a design for feasibility or safety. Analysis statements can be short or long, depending on the complexity of the analytical process. Analysis sections should answer simple questions about your work. These questions may include:

1. How do you know your experimental results are correct?
2. How do you know your design will be safe?
3. How do your results compare with benchmarks in your area?

Each of these questions speaks to a different type of project, but they all speak to the need for any technical professional to evaluate results dispassionately and thoroughly.

An analysis presentation can resemble a short experimental presentation, for it opens with a task definition, and it must account for the methods, including assumptions, models, and values that were used. It should also indicate areas where the problem has been adjusted for simplicity, where values have been selected in order to be conservative, or the like.

Our sample design report presents an extensive analysis in the section **Numerical Investigation of Proposed Splice Detail**.

Closing Summary

A workplace report should close with two small chunks of information: a point summary and a list of action items.

Summary

A closing section should first summarize the significant findings that have been mentioned elsewhere in the report. The task of this summary is simply to collect in one location those points that readers need to remember and that they may need to present to their superiors. These summary statements should be short; you should limit these to two lines of text per point, and you should consider formatting them as bullet points.

Action items

The second part of a closing section should specifically list any actions that you recommend that the client take, either to address problems from your point list or to schedule follow-up reviews at a later time.

Because our sample report fulfills a contract with an outside agency, there are no action items for our author to present. Consequently, our example closes with a summary only.

CHAPTER 3.3

EXAMPLE OF A DESIGN REPORT FOR A WORKPLACE PROJECT

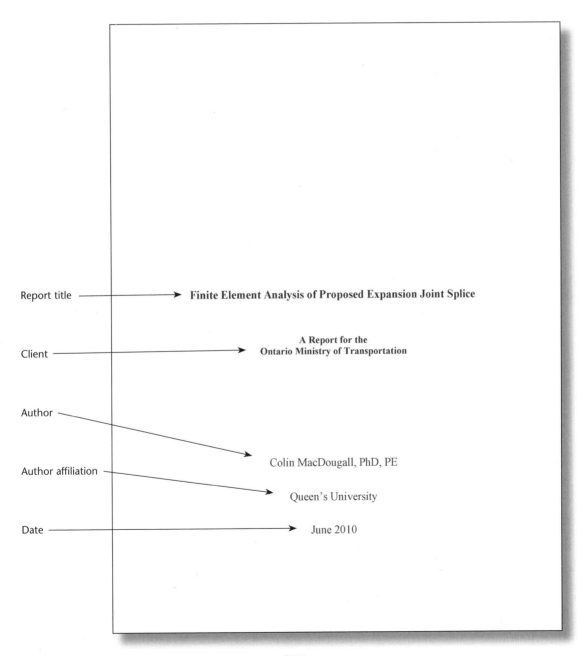

Report title → **Finite Element Analysis of Proposed Expansion Joint Splice**

Client → **A Report for the Ontario Ministry of Transportation**

Author →

Author affiliation →

Colin MacDougall, PhD, PE

Queen's University

Date → June 2010

Topic is defined for non-expert and expert readers.

Need is defined.

Related background information is presented.

Introduction

Modular bridge expansion joints (MBEJs), as in Fig.1, are proprietary components that provide allowance for expansion and contraction for bridges typically over 100 meters long. MBEJs are composed of several components, namely center-beams (CBs), support-beams (SBs), support boxes (which accommodate sliding SBs to provide the required movement), sealing strips, springs, and bearings. The components are typically made of galvanized steel and are connected together by welding. MBEJs are subject to considerable vertical and horizontal cyclic loads, making them susceptible to fatigue.

Centre-beam (CB)

Support box

Support-beam (SB)

Figure 1. Modular Bridge expansion joint (right picture adopted from [1]).

Premature failures due to fatigue have been reported in installed MBEJs [2 & 3]. These failures require either repair or replacement. Exhaustive research by Dexter et al. [4] examined the behaviour of MBEJs under static and cyclic loading. Among many issues needing more research, as stated in NCHRP report 467, is a fatigue-resistant field splice for centre-beams. To minimize traffic blockage, the replacement of failed MBEJs needs

1

Background
(continued)

Other splice
methods are ———————→
described.

Problem is
defined: poor ———————→
fatigue life of
splices.

to be performed in several stages, during which one or two lanes may be closed. The installation of an MBEJ will typically be staged, and therefore these joints need to be spliced in the field. It is not always possible to have optimum quality control over the splice installation due to the limited working space, as in Fig. 2, and to time constraints. The current designs for centre-beam field splices, whether of the hinge-type, as in Fig. 3, or the moment-resisting type, as in Fig. 4, do not possess the same fatigue life as do the other components of the MBEJ and thus cause premature, costly failures in the installed MBEJs. Chaallal et al., thus, recommend that "an improved splice be designed" (2002) [5].

Figure 2. View within a modular bridge expansion joint.

Goal defined. ———————→

Project scope:

• Requirements ———————→
• Design
  performance ———————→
  analysis

The Ministry of Transportation of Ontario (MTO) has requested Queen's University researchers to examine this problem. The study is to involve industry partners in Ontario and Quebec. This report presents a new design for a center-beam field splice. The next section presents the requirements for the new design. The splice design is then briefly presented, after which a numerical investigation is described in order to validate the performance of the proposed splice. The results are summarized at the end of this report.

2

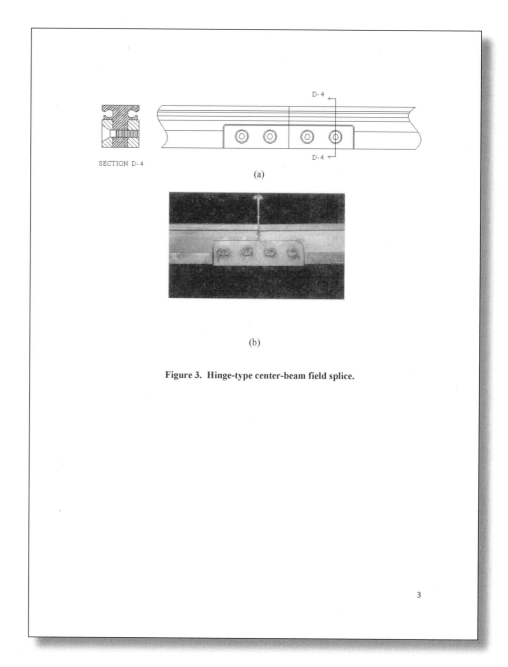

SECTION D-4

(a)

(b)

Figure 3. Hinge-type center-beam field splice.

3

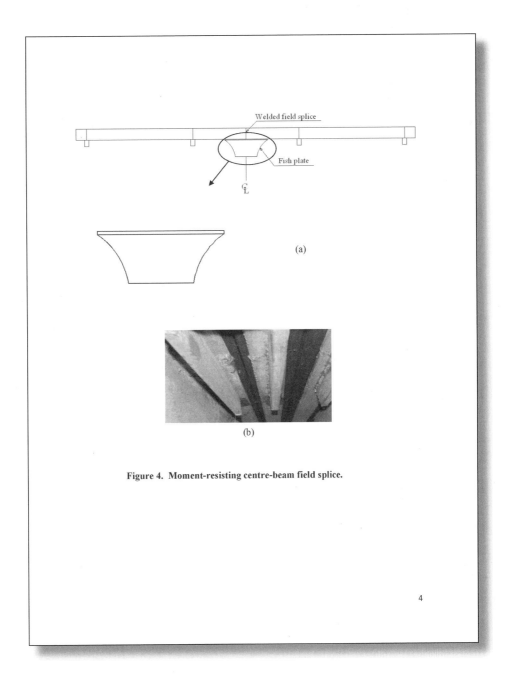

Figure 4. Moment-resisting centre-beam field splice.

4

Qualitative design criteria are listed. Specifications and other goals or requirements can be listed in this area as well.

Requirements for Fatigue-Resistant MBEJ Splice

A new fatigue resistant design for an MBEJ should meet the following criteria:

- The detail should be versatile enough to accommodate the variety of proprietary centre-beams on the market.

- The detail must be feasible and easy to install in the field.

- In order to be consistent with the fatigue resistance of the rest of the MBEJ system, the splice detail should be at least Category C [5].

- The detail must be resistant to all dynamic and static effects of both vertical and horizontal loads and must be durable over the design life of the MBEJ.

The design presented below is intended to be versatile and easy to install in the field. Finite element model predictions presented in the second half of this paper demonstrate that it is resistant to fatigue and to loading effects.

5

Design
description
with a figure:

- Solution is
  named and
  motivated.
- Figure is
  cited and
  identified.
- Labeled
  elements of
  figure are
  called out.

Proposed FRP Splice Detail

Fiber reinforced polymer (FRP) material has been suggested as an alternative to welded steel fish plates for splicing MBEJs and other steel components. Figure 5 shows the proposed splice for a single span centre-beam. An FRP box section would be epoxied to the bottom flange of the steel beam. The web of the centre-beam would be spliced using steel plates and pretensioned steel bolts to create a slip-critical connection.

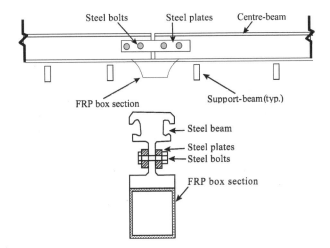

Figure 5. Proposed detail for centre-beam field splice.

6

An analytical
presentation
must define:

1) Methods and
   assumptions
2) Material values
3) Loads that must
   be accounted for.

1) Methods and
assignments are
defined first.

Finite Element Model Investigation of Proposed Splice Detail

In order to assess the feasibility of the proposed splice detail, a finite element model was developed and investigated. The finite element model was implemented in ABAQUS and is shown in Figure 6. Although the centre-beams of MBEJs are typically continuous over several spans, it was decided to model only a single simply-supported span to reduce the size and complexity of the model. In the future, the effects of continuity will be investigated. The span of the beam is 1 metre, which is typical of the centre-beams for an MBEJ, and pin and roller supports were assumed.

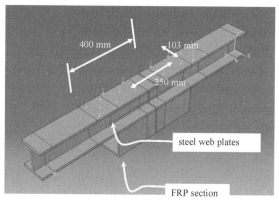

Figure 6. Spliced center-beam modelled using ABAQUS.

In developing the finite element model, 10-node modified quadratic tetrahedron (C3D10M) elements have been used to mesh the 3D model. The steel was modeled as a homogeneous material having E = 200 GPa and the FRP was modeled as an orthotropic material using the data provide by the supplier. The steel beam dimensions were selected to be similar to those of typical MBEJ centre-beams on the market (W100 × 19).

The FRP section was selected to have the same breadth as that of the center-beam flange. Both carbon fibre and glass fibre reinforced polymer could be used for this application,

7

although glass fibre reinforced polymer (GFRP) is cheaper and more widely available than CFRP.

2) Material properties and dimensions are specified, as are testing conditions.

Two different glass fibres (GFRP) with E_{11} = 17.1 GPa and E_{11} = 120 GPa and one carbon fibre (CFRP) with E_{11} = 400 GPa were investigated. The first GFRP, which has a very low modulus compared to steel, was selected because it is readily available; it was used for static tests of the detail. For the GFRP model, a wall thickness of 6 mm was assumed. For the CFRP model, a wall thickness of 2 mm was assumed.

The steel web splice plates were assumed to have cross-sectional dimensions of 6 mm × 60 mm. As per typical practice, there is a five-millimetre gap between the two halves of the steel beams.

3) Loads are described using diagrams of truck dimensions aligned with a diagram of wheel loads.

The loading on the model is based on the Canadian Highway Bridge Design Code CL-625-Ont truck loading [6]. This loading is typically used to design MBEJs. Fig. 7 shows the CL-625-ONT truck loads and spacing. Half of the axle load (that is, one wheel load) was applied to the splice in the form of two distributed loads, each resembling one of the two wheels comprising each side of the axle. The CL-625-ONT wheel patch of 600 mm x 250 mm was modified so that in the model the wheel load was applied on two areas of 250 × 103 mm with a spacing between the centers of 400 mm. These loads are indicated in Fig. 6 as well.

8

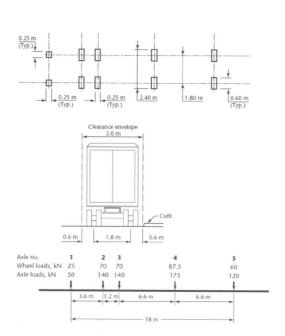

Figure A3.4.1
CL-625-ONT Truck
(See Clause A3.4.1.)

Figure 7. CL-625-ONT truck loading according to the
Canadian Highway Bridge Design Code [6].

It is assumed that the distance between adjacent centre-beams is such that each centre-beam will be subjected to the full axle loading. This is a conservative assumption. It is also assumed that pressure under each wheel is uniform.

9

266

4) Analytical goals are defined to establish: a) Stiffness and b) Fatigue criteria.

4a) Analysis for stiffness begins.

Results

The main purpose of the splice is to provide a moment-resisting connection for the ends of the centre-beams. Theoretically, the spliced beam should have the same stiffness as an unspliced centre-beam. In addition, the stresses within the spliced beam should be low enough to ensure fatigue does not occur.

The maximum deflection at the midspan was used as the stiffness measurement. The model was first used to determine the length of FRP required to ensure that the stiffness of the spliced beam is similar to that of an intact beam. In this case, a GFRP beam with $E_{11} = 120$ GPa was assumed. Table 1 shows the results.

Table 1. The effect of FRP section length on the maximum deflection of the spliced beam.

Beam Condition	Length of FRP section (mm)	Maximum Deflection (mm)
Intact	-	1.515
Spliced	200	1.982
Spliced	300	1.741
Spliced	400	1.484

Result of stiffness analysis.

The maximum deflection of the intact beam under the truck loading is 1.515 mm. The results indicate that the GFRP box beam ($E = 120$ GPa) must be at least 400 mm long to ensure that the maximum deflection of the spliced beam is similar to that of an intact beam. The typical span of a centre-beam is 1 metre, so a 400 mm FRP beam would still have 300 mm clearance on each side for installation in the field.

4b) Analysis for fatigue begins.

Fatigue analysis results:

Figure 8 is described.

Figure 8 shows the stress distribution of the spliced beam, in this case with an FRP box beam length of 400 mm. The spliced beam does not bend with uniform curvature, as the FRP box beam significantly stiffens the beam over the splice length. Regions of high stress are indentified in Figure 8: Location A is at the mid-span where the steel web splice plates meet the gap between the centre-beam ends; Location B is at the end tips of the

10

FRP section. The stresses at these locations are indicated in Table 2. The results in Table 2 are for 103 mm × 103 mm FRP box sections. Two different FRPs (glass FRP and carbon FRP) were considered, and the thickness *t* of each box section is indicated in the table.

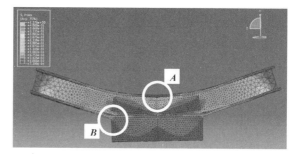

Figure 8. Stresses in spliced beam subjected to maximum truck loads.

Stress analysis results are presented in figures and tables. Table 2 is indexed to labels in Figure 8.

Table 2. Maximum stresses in splice with a GFRP or CFRP section. Locations *A* and *B* indicated in Figure 8.

	σ_{max} (MPa)	
	GFRP (*E*=120 GPa; *t*=6 mm)	**CFRP** (*E*=400 GPa; *t*=2 mm)
A	140	168
B	167	327

Analysis results are compared to expected stresses.

Steel will generally experience no fatigue for stress amplitudes below 50% of its static strength. Assuming 450 MPa for the tensile strength of the steel, the fatigue limit is 225

11

Modeling results are compared to calculated loads to validate the proposed splice.

MPa. The maximum predicted stress in the steel, 168 MPa, occurs at Location *A* when a CFRP section is used. This stress is 33 % below the fatigue limit, indicating that fatigue in the steel will not occur.

The fatigue stress-life curve for GFRP composites, when plotted on log-log axes, is linear with a slope of 10%, and it does not exhibit a fatigue limit. An equivalent fatigue limit of 25% of the tensile strength is suggested for GFRP [6]. For GFRP with a tensile strength of 700 MPa, a fatigue strength of 175 MPa can be expected. A fatigue limit of 65% of the tensile strength is suggested for CFRP [6]. For CFRP with a tensile strength of 1431 MPa, a fatigue strength of 930 MPa can be expected. Comparison with the stresses at Location *B* in Table 2 indicates that the predicted stress in the GFRP is 5% below the fatigue limit and the predicted stress in the CFRP is 65% below the fatigue limit. Although the GFRP stresses are close to the fatigue limit, it should be noted that a factor of safety is already included in the fatigue limit suggested in [6]. Therefore, this would be an acceptable fatigue resistant design.

12

Closing summary briefly restates the results that have been presented in the individual sections of the report.

Conclusions

An innovative method of splicing steel beams is investigated in this report. The splice detail uses bolted steel web plates and a fiber reinforced polymer box section epoxied to the beams. This detail was investigated for beams and loadings typically found in modular bridge expansion joints. Such a splice detail would be advantageous in this application because of its improved fatigue performance, which results from avoiding the use of welding.

A finite element model of the proposed splice detail was implemented and investigated. Glass fiber and carbon fiber were investigated as potential materials for the splice detail. The following conclusions result from the finite element analysis:

1) A minimum length of 400 mm for the fiber reinforced polymer section is needed to develop the required stiffness in the splice detail.

2) Glass fiber with a modulus of at least 120 GPa is needed to develop the required stiffness in the splice detail.

3) Both glass fiber and carbon fiber reinforced polymer can be used to develop a fatigue resistant splice detail.

13

References

[1] Hennegan and Associates. 2010. "Modular Expansion Joints." Electronic Product Brochure [online]. Obtained May 5, 2010. Available from henneganandassociates.com/expansion_joints.htm

[2] Roeder, C. W. (1998). "Fatigue and Dynamic Load Measurements on Modular Bridge Expansion Joints." *Construction and Building Materials.* 12(2-3), 143-150.

[3] Dexter, R. J., Osberg, C. B., and Mutziger, M. J. (2001). "Design, Specification, Installation, and Maintenance of Modular Bridge Expansion Joints." *Journal of Bridge Engineering.* 6(6), 529-538.

[4] NCHRP- Report 467. (2002). "Performance Testing for Modular Bridge Joint Systems." Edited by Dexter, R. J., Mutziger, M. J., and Osberg, C. B., National Cooperative Highway Research Program (NCHRP), Transportation Research Board (US), Washington, DC.

[5] Chaallal, O., Sieprawski, G., and Guizani, L. (2006). "Fatigue Performance of Modular Expansion Joints for Bridges." *Canadian Journal of Civil Engineering.* 33, 921-932.

[6] Can/CSA-S6-06. (2006). "Canadian Highway Bridge Design Code." Mississauga, ON; Cl. 3.8.3.2 and Annex A3.4.

14

CHAPTER 3.4

GUIDE TO PRESENTATIONS IN THE WORKPLACE

Introduction

Earlier in this book we offered simple guidelines for presenting experimental results to teachers. Those classroom presentations resemble tests; the instructors examine displays in search of mistakes, and the students are not usually asked to motivate their projects or to recommend actions. Workplace presentations are different; speakers usually are in a position to recommend that actions be taken, and audience members expect speakers to tell them things they did not previously know. Other factors impact workplace presentations as well; audiences do not necessarily know anything about your subject before you begin speaking, and they are often pressed for time. These audience members want you to use their time efficiently. This means they want you to inform them quickly about your topic, they want you to sharply define any actions that you want them to take, and they want you to give them enough information to feel that those actions are appropriate and reasonable. In this chapter we will give some general tips for designing presentations that are brief and that speak efficiently to the non-expert members of your audience. We will begin by outlining some strategies for thinking about time and audiences, and we will explain how those strategies govern the organization of a typical technical presentation. We will present and discuss a brief sample presentation, and we will finally explain how to avoid some common and simple problems of slide design.

Planning Your Presentation

Time

The key to success in a presentation is to recognize that you must be brief. In the (usually short) time allotted, you will be unable to give a comprehensive discussion of your work, and you should not attempt to do so. Rather, you should plan to display a small number of results and to make a small number of points during your time, and you should look for opportunities to present additional details during a question-and-answer period following the presentation.

Your results, of course, are displayed on slides as diagrams, plots, or tables, and you will require a certain amount of time to describe the images on these slides, usually around one minute per slide. Several factors impact the amount of time you should allow for presenting slides. Plots and tables, for example, can often be described relatively quickly, as they tend to display numerical results that can be framed concisely. Drawings and diagrams often display more details and raise practical concerns about orientation and about integration with existing systems; these concerns can force a speaker to spend more time describing such images than they might spend describing a plot or table. Consequently, speakers whose results are presented as drawings must sharply limit the number of slides they will present; a speaker whose results are primarily captured in tables commonly has more flexibility in managing time during a talk.

Goals

In order to plan an effective workplace presentation, you need to set goals for your audience and for your project. That is, in order to assemble details effectively, you need to decide what action you want the audience to take after you finish speaking and how that action should impact the work you and your team are doing. The goal statement need not be elaborate; you should be able to state it in a simple sentence, such as one of the following:

- Your audience should approve your request for funding.
- Your audience should approve your request for additional time on the project.
- Your audience should understand a new job assignment that you've given them, and they should know what steps to take first.

You will set different goals for different stages of your projects, of course. Your goals may also vary according to your seniority; the goal statements above make requests and assignments that are more appropriate for a project leader than for a new member of a project team.

Presentations do not always involve funding requests or job assignments. Many workplace presentations simply disclose information that audience members need to have in order to make plans and to meet deadlines. These presentations can be broadly described as project status reports and administrative policy statements. Project status reports integrate a review of a project's schedule with a summary of current accomplishments, problems, and plans. This information allows team members to make adjustments to their work in order to accommodate progress on other sections of the project. There are often questions at status report presentations, as team members try to determine how their tasks are impacted by progress on other tasks.

Another type of information disclosure is the administrative policy presentation. Here a speaker explains adjustments that the company is making to the rules—governing matters ranging from network security to the reporting of travel expenses—that employees must follow. Policy presentations need to present large action descriptions for the audience, as listeners will need to understand what steps they must take in order to successfully comply with new rules.

Your audience

It is hard to make generalizations about workplace audiences, as your listeners have different amounts of seniority and different kinds of training. However, you can make two simple and powerful assumptions about colleagues and supervisors who are not members of your daily task team:

1. Your listeners are not paid to know all that you know about your task.
2. Your listeners cannot remember everything you have explained to them in previous presentations.

Your colleagues generally cannot track the details of your project tasks while also making progress on their own tasks. They are too busy to do this successfully. Rather, they will rely on you to give them a clear overview of your task in each presentation and to follow that overview with details describing the interface between their tasks and yours. Consequently your audience analysis should avoid efforts to define what your colleagues might remember about your work before the presentation starts. Rather, you should define what colleagues should do when your presentation has finished. Generally this means you should organize your presentation to describe your project goals, your current status, your completion schedule, and the way your work will interface with the work of your colleagues.

Presentation Organization

A presentation should be viewed as an oral-visual summary of a written report. Consequently, it should be organized according to the same plan that is used in a written report, with some adjustments to accommodate time and audience constraints—it should present less detail than is found in a written report. To respond to audience needs while summarizing a full written report, a presentation should be subdivided into sections as follows:

1. Cover slide
2. Outline
3. Introduction
4. Results
5. Discussion
6. Recommendation and Conclusions
7. Questions

This outline and the guidelines below are designed to describe a presentation that is ten minutes long. The example report that is presented in the next section is also designed to represent a ten-minute presentation.

Cover slide

The **Cover Slide** of a presentation does the same job as the cover sheet of a report—it identifies the presentation's topic, the author(s), the author's professional affiliation, and the date of the presentation. This information requires no comment from the speaker, but because it identifies the speaker, the topic, and the occasion, it should be on display when the speaker is introduced.

Outline

An **Outline** is the Table of Contents for a presentation; it shows the topics you will introduce and it shows how these are related. In presentations of 10 minutes and longer, you should display an outline as you introduce your topic.

Introduction

The **Introduction** section of your presentation should be broken into two chunks, and it should fit on two or three display slides. The first slide should present a Background statement, which should name the topic of the presentation and the problem you are addressing. When you are working for a client, this section should recite the client's needs and constraints while listing any other details that may be pertinent to your project work, such as the age of a machine that you must fix, the size of a component that you cannot remove, and the like. The second slide should motivate the work by explaining why the problem is important and by explaining the consequences of inaction. The last slide of your Introduction can be called a goal statement or an objective statement; in this slide you should specifically state the actions you are taking to address the problem, and you should indicate what results you will describe during your presentation.

In our sample presentation, the Introductory information fills five slides, using a Cover Page, a single Outline slide, and three slides for an Introduction section.

Results

The second large section of a presentation describes your accomplishments on your project. This section of your talk presents either the **Results** of your work on a problem or, if the work is ongoing, it should present the **Status** of your task, describing work that has been completed to date and comparing this work to your schedule for completion.

Regardless of whether your task is complete or not, you should rely on figures such as drawings, diagrams, or plots to represent your work. If you are delivering a status report on a large task, you should present illustrations to represent any work that has been completed to that point, and you should augment these with displays of your project schedule in order to demonstrate how effectively you are meeting the schedule for your task.

In our sample ten-minute presentation, the Results are presented using only two slides.

Discussion

The third large section of a presentation can loosely be called the **Discussion.** Under this heading you should offer explanations or warnings that are significant to the audience's understanding of the results that you have presented. These explanations might include descriptions of alternative solutions that have been rejected, analyses of error or of safety, evaluations of your project's schedule, and the like. The Discussion section of a talk should be brief; you should focus on only two or three significant concerns that your work might raise. Other concerns can be deferred to the question and answer session that follows the formal portion of your presentation.

The size of a Discussion section can be deceptive. The Discussion section of our example report is limited to two tables, shown on two slides, yet these tables represent a significant amount of analytical work.

Recommendation

The next section of a presentation can be called the **Recommendation** section. Recommendations can be presented in a variety of ways, according to the project. When tasks are to be distributed among team members, recommendations are presented as *Action Items*. For the design or purchase of components, recommendations can be presented as *Specifications*.

For many projects, the recommendations can be presented in a statement of a single line. When this happens, it is reasonable to combine the recommendation slide with a Closing or Summary slide, as is the case in our example presentation.

Summary

The last section of a presentation is a Closing or Summary. This should be presented as a single slide that aggregates the most important points from your Results and Discussion sections. It is common for Summary slides to combine numerical data with action recommendations.

In the example presentation below, the closing Summary is confined to a single slide.

Questions

The last statements of any presentation should be these: "Thank you, I will be happy to answer questions." These statements establish that your talk has concluded, and they announce the beginning of the unstructured Question session. Circumstances will dictate the length of the question session, although these tend to be brief.

While the question session is unstructured, it presents you with an opportunity to present information that you could not fit into the main section of the talk. To take advantage of this opportunity, you should prepare slides to present any data and analysis points that you would like to explain further, and you should prepare responses to questions that you think your formal presentation might raise.

In responding to questions, you should do two things:

1. Repeat questions after they are asked. This helps other audience members to hear the questions, and it buys you time to organize your response.
2. Locate and display pertinent slides before you begin your answer.

When you do these things consistently during the question session, you will help your audience members by giving effective answers, and you will help yourself by controlling the tempo of the discussion.

CHAPTER 3.5

EXAMPLE OF A DESIGN PRESENTATION FOR A WORKPLACE PROJECT

Introduction

The following is a typical engineering presentation. It describes the results of a stress analysis performed on a component. The presentation is divided into the following sections:

1. Introduction
    Background
    Goal
2. Results
3. Discussion
4. Closing Summary

In addition to these information sections, the presentation includes a Cover Slide and an Outline slide, both of which are presented before the Introduction slides.

This presentation was limited to 10 minutes with time for questions after the formal presentation had concluded. You will note there is no Recommendations section. This is because this presentation described a completed analysis project; there was no follow-up work planned.

Overview of Sample Presentation

Section 1: Title and Organization of the Presentation
Two Slides

Fiber Reinforced Polymer Splice
for Expansion Joints

Colin MacDougall, PhD, PEng

Queen's
UNIVERSITY
January 20, 2010

1. Cover slide.

Outline

- Introduction to Expansion Joint Splices
- Finite Element Model
- Stress Results
- Discussion of Size and Material
- Conclusion

2. Outline.

Section 2: Introduction
Three Slides

Expansion Joints
- Allow for expansion of bridge due to temperature
- Subject to severe fatigue loading due to truck loads

1. Background.

Splicing Beams
- Need to connect beams in expansion joints
- Welding beams can lead to fatigue cracks

2. Problem.

New Splice Design

- Company proposing to splice beams using a glass fiber reinforced polymer (GFRP) box
- Avoids welds
- GFRP is fatigue resistant
- Require stress analysis

3. Goals and challenges.

Section 3: Results
Two Slides

Finite Element Model

- 10-node modified quadratic tetrahedron (C3D10M) elements
- Pressure loads to represent truck wheels
- GFRP modeled as an orthotropic material

1. Modeling approach.

Results - Stresses

- Large stresses at GFRP to steel connection
- Potential location for fatigue cracks

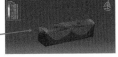

2. Results.

Section 4: Discussion
Two Slides

Effect of GFRP Length

- GFRP must be at least 400 mm long to ensure adequate stiffness

Beam		δ_{max} (mm)
Intact beam		1.515
Spliced beam (using GFRP E= 120 GPa)	$L_s = 200$ mm	1.982
	$L_s = 300$ mm	1.741
	$L_s = 400$ mm	1.484

1. Analysis of length.

Effect of FRP Type

- Using glass fiber rather than carbon fiber results in lower stresses

Type of FRP vs. max. stress developed in the splice			
σ_{max} (MPa)		GFRP (E=120; t=6; L_s=400)	CFRP (E=400; t=2; L_s=400)
Region of splice where max. Stress (σ) is calculated	Midspan	140	168
	End of FRP section at its junction with steel beam	167	327

2. Analysis of type.

Section 5: Closing Summary
One Slide

Conclusions

- Large stresses at GFRP to steel connection can occur

- GFRP box must be at least 400 mm long

- GFRP will result in lower stresses than CFRP

Planning Your Slides

Before preparing your slides in detail, it is important to prepare an outline of the presentation:

1. Start by defining how much time you have to speak. This will allow you to determine approximately how many slides to use. *A good rule of thumb is "one slide per minute."*

2. Once the total number of slides is defined, determine the number of slides available for each of the major sections of the presentation. Usually three slides will be taken up for the Title, Outline, and Closing Summary. The remaining slides need to be proportioned between the Introduction, Results, Discussion, and Recommendations. *A common mistake is to create far too many slides and spend far too much time on the Introduction section of a presentation.*

3. Once the number of slides available for each section has been defined, you can identify the critical ideas you want to convey to your audience on each slide. *Start by identifying the two or three critical results.* Decide how you want to display those results (usually a figure, graph, or table). Do the same for the Discussion and Recommendations. Finally, outline the slides for the Introduction and Closing Summary. *Don't worry at this point about how the slides look.*

4. Once you have an outline of the presentation, run through the presentation from start to finish to ensure the ideas flow logically.

5. When you are satisfied with the outline of the presentation, you can then work on formatting each slide so that it clearly presents its information to the audience.

Title

The Title slide provides the title of the presentation, the name and credentials of the presenter, the affiliation of the presenter, and the date of the presentation.

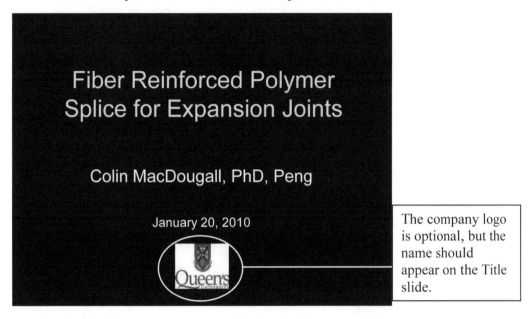

Often there is more than one contributor to the work that is being described by the presentation. In this case, all contributors are usually listed on the title slide, with the presenter listed first:

Fiber Reinforced Polymer Splice
for Expansion Joints

Colin MacDougall, PhD, PEng, Queen's University

Dev Alahan, M.Sc., Queen's University

Peter Barlow, B. Sc., BridgeJointCo Inc.

January 20, 2010

Outline

The Outline provides the main headings for the presentation. There should rarely be more than 5 or 6 main headings for the presentation. When presenting the Outline slide, do not go into details. This wastes time, and the details will come later in the presentation.

Outline

- **Introduction to Expansion Joint Splices**

- **Finite Element Model**

- **Stress Results**

- **Discussion of Size and Material**

- **Conclusion**

While you should not give details in your outline, you should give specific information about your presentation. You can do this efficiently by showing topics for each of the general section headings. Here, for example, the vague heading "Results" is changed to the specific "Stress Results," and the vague "Discussion" heading is modified to add the specific topics of Size and Material.

Introduction

The purpose of the Introduction is to help the audience appreciate why this problem is important (Background), and to define the particular aspect of the problem you are investigating (Goal). The Introduction is also an opportunity to define terms that will be used later in the presentation.

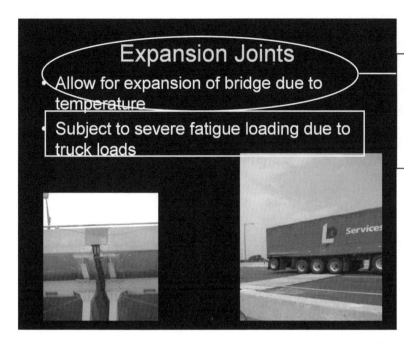

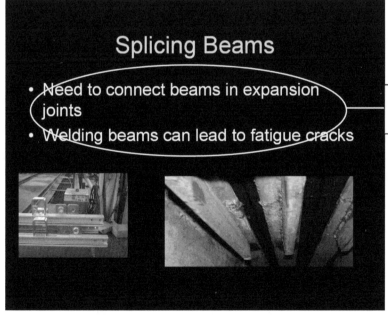

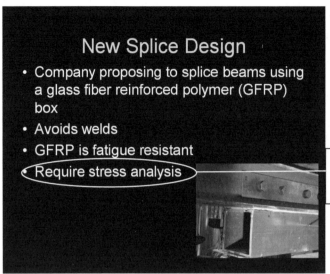

An Introduction must be efficient in order to save time. To introduce your work efficiently, you should concentrate on four definitions that the audience must see:

1. The topic
2. The problem
3. The importance of the problem
4. The task you performed to address the problem

Information that does not serve these definitions is likely to distract your audience.

Results

You should begin a Results presentation by describing the method used to obtain the results. You will likely not have enough time to provide all the details on the methods during the presentation. Often a report is written in conjunction with the presentation. The report should contain detailed information on the Methods. A good tip to save time is to state that: "Detailed information on the methodology has been provided in the final report." Of course, be prepared to answer questions if they arise (it is often worthwhile to have additional slides prepared to help answer these questions).

You will likely not have enough time to present all your results during your presentation. Therefore, narrow your focus to the two or three most important results.

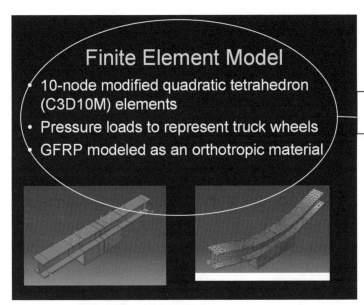

Finite Element Model

- 10-node modified quadratic tetrahedron (C3D10M) elements
- Pressure loads to represent truck wheels
- GFRP modeled as an orthotropic material

Describes the method used to investigate the problem.

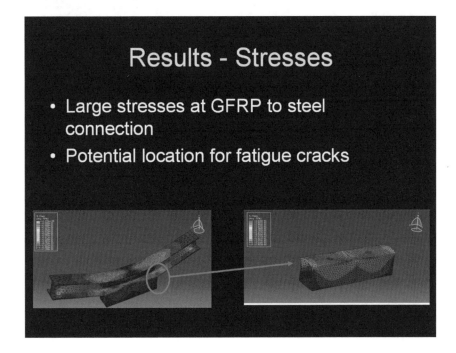

Results - Stresses

- Large stresses at GFRP to steel connection
- Potential location for fatigue cracks

Discussion

For this presentation, the Discussion involved two tables. Tables often contain too much information for the audience to absorb. In the slides below, red boxes were imposed on the tables to draw the audience's attention to the most important points in each table.

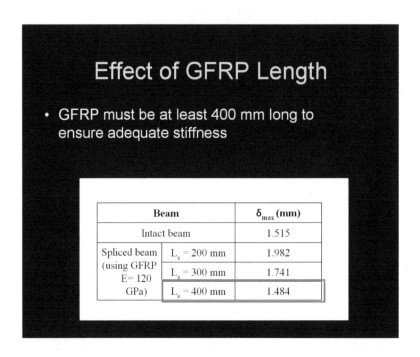

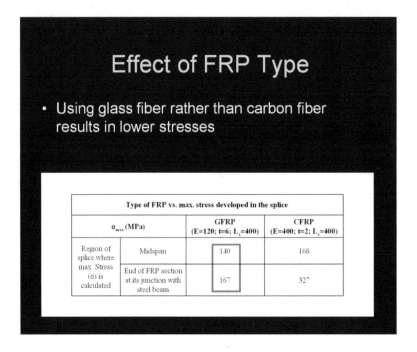

CLOSING SUMMARY

In the Closing Summary, display again the most important two or three points from the presentation. *Do not introduce any new information not discussed during the presentation.*

287

Chapter 3.6

Guide to Performance During Presentations

Introduction

Presentations differ from written documents in that you are on display for your audience at all times, and your actions can have an impact on the audience's response to your words. There are several things that you can do that will assure that your audience understands what you are saying:

1. Rehearse the presentation.

You should speak through your presentation fully before you appear before your audience. It is best to rehearse in the room where your talk is scheduled, as this will allow you to get accustomed to the projection equipment that you will use. If you have no access to that room, however, you can simulate the presentation by advancing slides on your own computer. This will help you to determine specifically how you will explain your figures and tables, and it will help you to orchestrate the small steps involved with advancing slides. If you are presenting with a partner, you should rehearse with that person; it is important for each member of the team to hear what the other will say, and it is important to know how much time your partner needs in order to complete his or her section.

2. Make points as you speak.

Your slides present information that the audience needs to see and remember. Audiences understand presentations best when you can state the takeaway information—the point—of each slide in a single sentence. You should develop that point sentence as you prepare each slide. An easy way to do this is to complete this sentence in your notes for each slide: "From this slide, my audience needs to remember [*blank*]." The statement in the bracket should fill one simple sentence. If the point is longer than one sentence, then your slide may be cluttered with unrelated ideas.

3. Use a pointer.

When you describe drawings or tables, you should direct the audience's attention to particular parts of those displays while you describe them. For this you need a pointer. When you point at the screen, keep your pointer still; when you move your pointer in large circles, audience members tend to lose focus on the components you are describing.

4. Face the audience as you speak.

You will naturally look at your displays as you move your pointer, and you will look at your projector as you change slides. However, you need to make sure that you turn your head toward the audience during most of the time when you discuss each slide.

5. Prepare handouts.

When you give a presentation, you should prepare a handout to give your audience. They might use this to take notes or they might review it later to remind themselves of your points. Speakers often distribute copies of their slides as handouts. This is acceptable, but it may not provide your audiences with adequate information, particularly if you have followed our advice to sharply limit the words on your slides while relying on figures to present your information. You should also prepare a one-page handout that captures the points and explanations that you have omitted from your slides, while aggregating your two or three most significant results displays.

The Logistics of Slide Presentations

Even at small businesses, people are expected to use computer projection systems to give presentations, so you need to understand how these systems work before you prepare your talk. While we cannot predict the particular set of plugs and buttons that you will encounter each day, we can outline some of the constraints that business projection systems impose on users so that you develop a default presentation that is robust to the most common problems. Here, we speak not about the computer systems so much as we speak about the projectors and screens that actually display the information you load into your computers. We begin by speaking generally about the problems of screen space and light. After this we will discuss the problems of slide design and the way you can use slide space to clarify information and to acknowledge your affiliations.

Screen space

Most business projectors use a 4 : 3 aspect ratio. This means that they cast light in a rectangle that is wider than it is tall. Narrowly, this means that you should choose landscape orientation when you review the settings for presentation slides or pages. Additionally, because the letter-size page to which many programs default does not have 4 : 3 proportions, projectors may resize images in a way that degrades your display. PowerPoint solves this problem automatically, as its default slide, measuring 10 inches × 7.5 inches, uses this 4 : 3 ratio. If you design your presentations using other programs, you will need to verify that your page settings will fit a similar 4 : 3 frame without distortion.

Light

Presentations depend on light, and light can be unreliable. The source of light for your presentation is a projector's lamp; your eyes then receive that light by reflection from a screen. The lamp and the screen can both be sources of distortion for your displays; projector lamps fade as they age, and screens lose reflectivity as they accumulate dust. When this happens, colors can fade and distinctions can be obscured. Light infiltration from windows or open doors can create additional distortions that are hard to control.

While you cannot control the projection equipment that you will use, you can plan for worst-case situations by developing slides that display well in poor light conditions. Usually this means reducing the amount of color you use in your slides; to the extent possible, you should exchange dark or patterned backgrounds for light, solid colors and exchange color illustrations for black-and-white line drawings. You should not try to eliminate color from your slides; however, you should select colors in such a way that your information will display well even if the colors fade toward gray. In practical terms, this means that line drawings tend to display more reliably than solid model drawings, and that photographs usually display poorly unless they are heavily annotated or edited. If you wish to edit a photograph to enhance contrast for display, you must disclose your editing steps to the audience, usually by displaying the unedited image before presenting an enhanced version.

Responsibility

When you give a presentation, you should always display a cover sheet showing your name, the names of your team members, the title of the work, and the date of the presentation. The presentation date should be visible on your slides as well. These displays allow your immediate audience to spell your name correctly, of course; more importantly, however, these define who is responsible for the accuracy of the information that is being presented. As a professional, it is important for you to do this because your slides are likely to be transmitted to people who did not attend the presentation, and they may be reviewed years after you have left the project. The people named on the slides should be ready both to receive credit when work is well done and to answer questions should problems arise later.

You should also list the names of your company and/or your sponsor on the cover slide, either with a text entry or with a corporate logo. Such listings provide a professional acknowledgement for those who may have funded your work, and they disclose the parties who may claim ownership of any intellectual property you may have created.

Slide geography

Presentation programs such as PowerPoint use a default slide size of 10×7.5 inches. When your projector and screen are placed correctly, such default slides will exactly fill your screen. However, slides are typically subdivided into a central display block and a border, as suggested in Figure 1.

In Figure 1, the light-colored areas at the center represent space that you can use to display your information. The shaded areas at the perimeter of the slide represent space that is commonly reserved for identity information—authorship, date, team, and/or corporate emblems

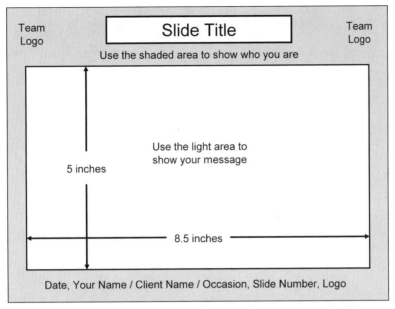

Figure 1. Space arrangement of a typical 10 × 7.5 inch slide

and the like. Some companies will require you to use a pre-defined company slide that has a wide border for corporate emblems; other companies will give you freedom to enlarge the display space while minimizing the corporate branding displays.

Displaying information

In a presentation, you provide two kinds of information—words and images. Words can be displayed in the form of text slides, labels on drawings, and column heads in tables, while images include drawings, diagrams, plots, and photographs. Usually your slides will present a mixture of words and images, and your challenge is to design that mixture in a way that is clear and that displays effectively. Handy rules of thumb are these:

1. Use type fonts between 20 and 40 point.

In most rooms, audience members will be unable to see words that are smaller than 20 point in size. This size restriction limits the amount of text you can display on a screen; you should avoid complete sentences in favor of key-word reminders for you. If your points require extended text discussion, you should place that discussion in a handout that audience members can review later. During the talk, it is better for them to listen to your explanation than to read your slides.

Figure 2 presents an example slide that uses 40-point Arial type throughout.

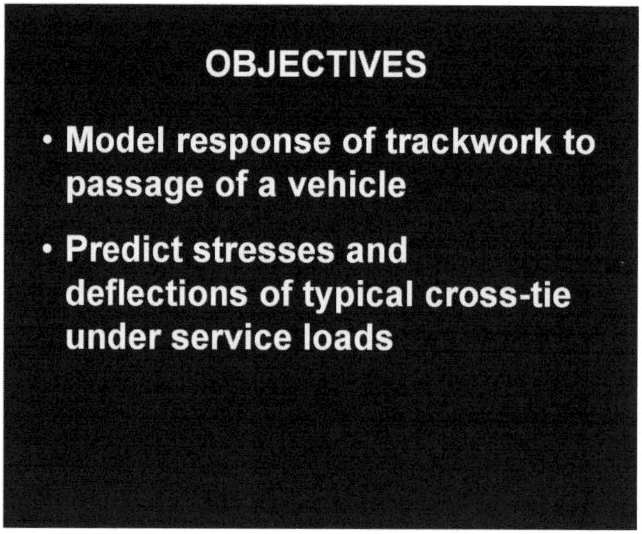

OBJECTIVES
• Model response of trackwork to passage of a vehicle
• Predict stresses and deflections of typical cross-tie under service loads

Figure 2. Text slide with 40 point type

2. Design images to use all of the available display area.
In presentations, plots, tables, and drawings should be large. You should prepare your visual information so that it will fill the display space on your screen as in Figure 3. Images are used to convey information that cannot be conveyed efficiently in words; they capture something crucial about your work, and you need to make sure that your images are large enough to display that crucial information sharply.

In Figure 3, even though the components are well labeled, they occupy only a small portion of the image, and their color differences are small. This image needs to be large in order for the bolt, plate, and pad to be easily visible.

HSS
Anchor Bolt
Gusset Plate
Elastomeric Pad

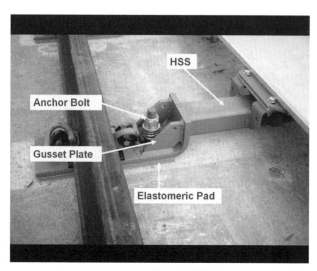

Figure 3. Images should fill the screen

When you display an image in an oral presentation, as in a written report, it is important for you to speak to the figures in detail. To do this, you should

1. Introduce the figure by specifically naming the device or component that is on the screen.
2. State why the figure is being shown by defining the point it makes or the result it presents.
3. Call out the labeled components of the display.
4. Discuss the display, explaining how it works or how you will use it in your work.

3. Minimize words on slides with figures.

Speakers commonly mix words with figures. This is a matter to approach with caution; in some circumstances words can enhance a figure display, while in other circumstances they create more problems than they solve. In general, words can usefully be applied to an image in the form of slide titles or labels for important features of the image. Such labels should be descriptive, short, and they should be placed near the features they describe; slide titles should naturally fill the border space at the top of the slide. Figure 4 is an example:

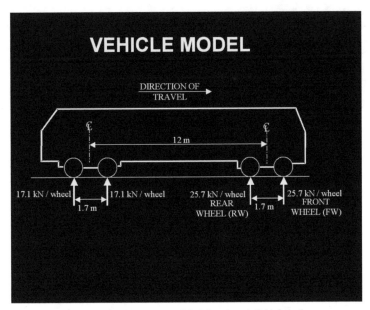

Figure 4. Diagram with clearly visible labels

Text creates problems for you when it is used to explain or comment on an image that is already labeled. Discursive statements are often lengthy, and they occupy valuable slide space; they may well obscure the components that are being evaluated.

Figure 5 is an example of a poor mixture of text and image, as the words and the image occupy roughly equal amounts of slide space. This is unacceptable; the text block uses a large amount of space to present trivial information, while the plot is so compressed that distinctions between the two data sets are rendered invisible.

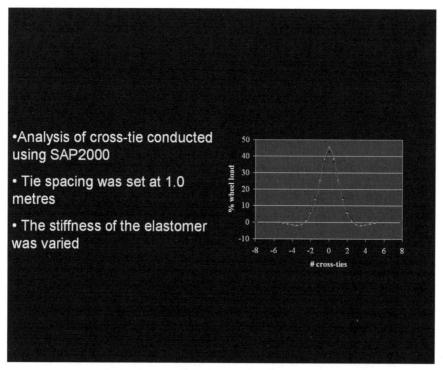

Figure 5. Side-placed text reduces space for displaying images

If commentary must be displayed on an image slide, it must be short, and it must be placed on an unused area of the slide, such as a border area or, in the case of an irregularly shaped image, on an area that remains vacant after the image has been sized. In Figure 6 below, the plot has been resized to fill the slide, and the side-placed text has been removed. Labels are used to convey critical information that is often presented in text bullets. Because images are commonly wider than they are tall, some text information can usually be placed above or below your display in the form of a slide title or a legend.

In this configuration distinctions between the data sets become visible, and this slide can stand on its own if circulated by members of your audience.

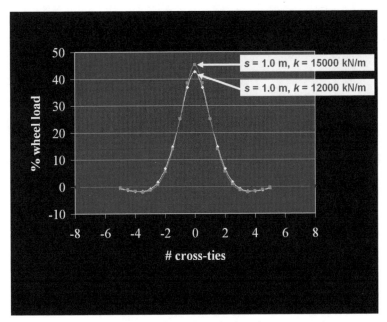

Figure 6. With text removed, the plot is resized
Critical information is inserted in label form.

Slide Designs and Backgrounds

Presentation programs such as PowerPoint make it easy for users to design slides with richly colored backgrounds. These programs come with libraries of slide design schemes and with tools that allow users to develop and share alternative backgrounds. Some of these backgrounds support good presentations and others do not. Our purpose here is not to impose a particular design and color scheme on your slides but to outline the general issues you should consider as you begin to develop your own presentations or to develop a template that your company might use widely. To do this we will present several commonly used slide designs, and we will discuss the strengths of each and the problems that they present.

Color

Slides communicate information to audience members' eyes. Because light conditions can be poor, it is best to prepare slides with high contrast between the background and the information that you want your audience to see and remember. The simplest way to obtain high contrast slides is to place dark text and lines on a white background. Black-on-white slides are easy to prepare, and they are of sufficiently high contrast to display effectively even under poor lighting conditions. Such slides also reproduce well as hard copies; this is an important matter, as presentation slides are often printed or copied using black-and-white office machines. The example black-and-white slide in Figure 7 was first introduced in an earlier chapter:

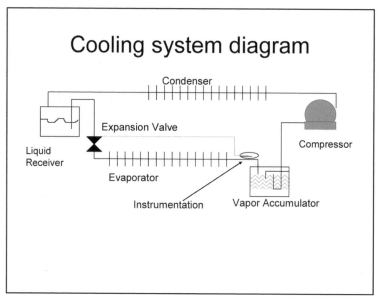

Figure 7. Black-on-white slides display well, but many authors find these boring

While black-on-white slides are excellent for reproduction, many authors find them to be dull, and some audiences find white backgrounds to be uncomfortably bright. To address these problems, many professionals invert their color schemes, using blue backgrounds for light-on-dark displays. When slides mainly present text, such backgrounds work nicely; however, diagrams and drawings tend not to display well on dark backgrounds, as indicated in Figure 8.

Here we have reproduced the cooling system diagram from the previous slide, and the slide background here clearly obscures the diagram. One can, of course, re-color the lines in this diagram in order for them to display well against the dark background, as in Figure 9.

Here the author has altered the technical display in order to accommodate the dark background, and this is a mistake. In addition to taking on unnecessary work, the author has trivialized the substance of the presentation by making it serve the presentation's color scheme. When you have the opportunity to choose, you should adjust the slide background rather than your professional output. This will save time for you, and it will make it easier for you to reproduce and adjust your slides later.

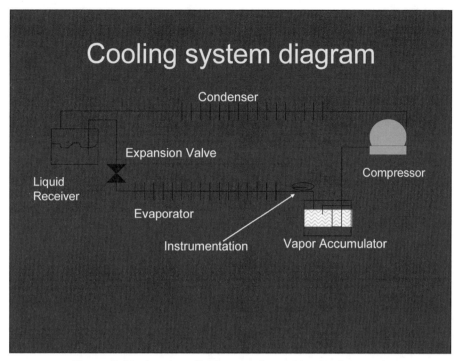

Figure 8. Dark colors display poorly on blue backgrounds

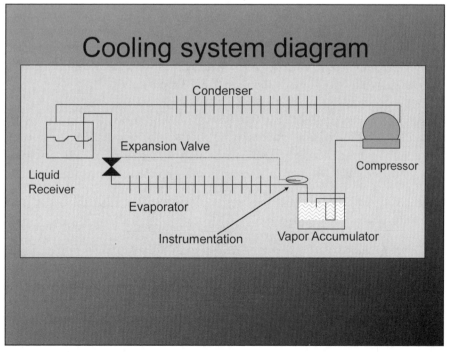

Figure 9. A diagram is revised to accommodate a slide's background color

You can enhance a presentation without distorting your information. Presentation programs will allow you to develop your own background colors quickly and easily. Generally speaking the most robust backgrounds will be light in hue and intensity and high in brightness. The example slide in Figure 10 shows a background that was developed using this principle:

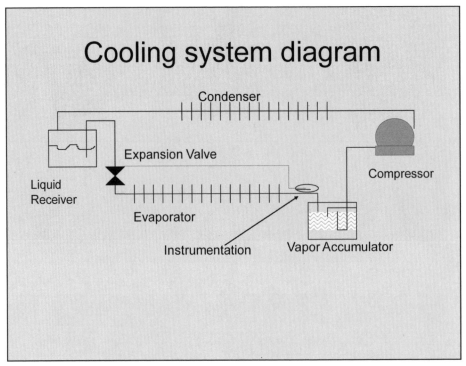

Figure 10. Robust slide backgrounds should be light in hue intensity and high in brightness

The slide is still blue in hue, but the color intensity is low, and the color is relatively bright. By adjusting these two factors, you will be able to explore a wide range of background colors, many of which will give you satisfactory results as presentation backgrounds.

Patterns

Many authors avoid solid-color backgrounds in favor of patterns, explaining that solid colors leave distracting empty space on slides. Presentation programs offer such authors patterned and decorative backgrounds to aid in the visual design of their slides. However, as was the case with color backgrounds, patterned slide backgrounds can present as many problems as they solve, and they should be used with some care. Figure 11 is an example of a blue slide to which a gradient color effect has been added such that the slide darkens in a diagonal from the top left to the bottom right.

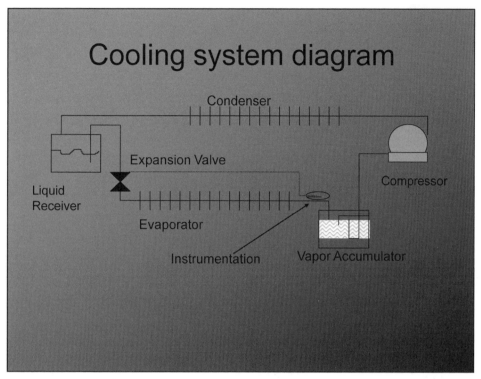

Figure 11. Gradient color schemes can generate an array of contrast problems

While gradient slides such as this one are commonly used, they can compound the problem of color contrast that we discussed earlier. With a gradient background, authors must work harder, because they must review the contrast that their displays form with several distinct color areas on their slides.

Background images

As an alternative to gradient backgrounds, authors may select from a variety of pre-bundled background images. One of these is shown in Figure 12 as part of a slide that again displays our cooling system diagram; it should be clear that several things are amiss. The background image is unrelated to the diagram being presented, and this creates a distraction; if the audience is paying attention, they are likely to expend some energy considering this disconnect. More generally, there is some risk that the background illustration could be more interesting than the diagram presented on the slide, a distraction that you as a presenter do not need to have.

Technical professionals can always take steps to mitigate the problems raised by their backgrounds. In Figure 13, the color problems of the gradient background are addressed by creating a secondary background—a drawing box—for the image. While this trick solves the display problem raised by the decorative background, it does so by resorting to the kind of solid, light display background that was described earlier.

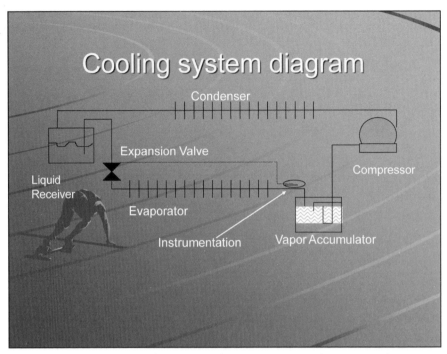

Figure 12. Decorative schemes can be more interesting than your information

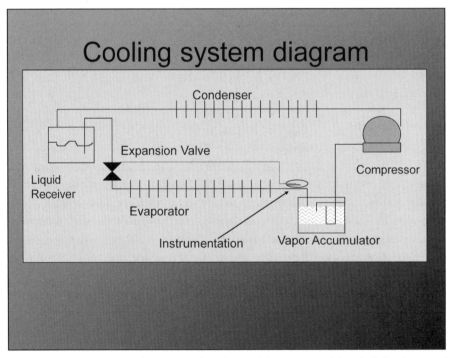

Figure 13. A light display box can solve the problems created by dark backgrounds

Space

Slide design schemes present authors with numerous problems regarding reasonable use of color and decoration. However, design schemes can also impact the space that is available to you to display your information. Figure 14 shows a widely-used slide design scheme that places a decorative, non-intrusive splash of color in the upper left side, along with a long, horizontal bar near the bottom of the color design. That horizontal bar defines the bottom edge of the slide title box, setting aside the top two inches of the slide as a title box. The slide design is pleasant, but the decoration fills up a great deal of space that could be used to display your figures.

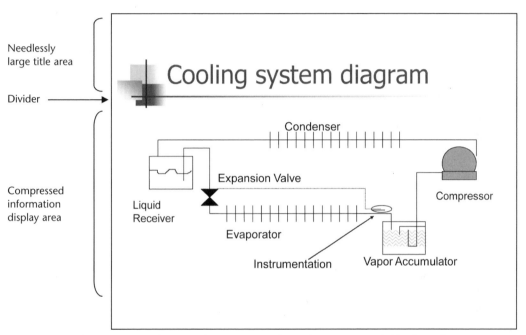

Figure 14. Decorative design schemes can reduce the space available for information

Some slide designs impact horizontal space as well as vertical space, as in Figure 15. As did the design in Figure 14, this slide decoration devotes two inches to the slide title, and it further imposes a needlessly wide border on the left side of the slide. Our cooling system diagram must be resized in order to fit it onto a slide using this design.

You can, of course, move these horizontal bars and borders by editing the master slide, and you should do so if you wish to display technical information on slides using these patterns. Alternatively, you may select slide designs that restrict decorations to the border area that was described earlier—slide space that you usually keep empty to make room for corporate emblems, and to accommodate misalignment of screens and projectors. In Figure 16, a decorative frame has been imposed just at the edges of the display area. This offers a pleasing band of color while not disrupting the display of information.

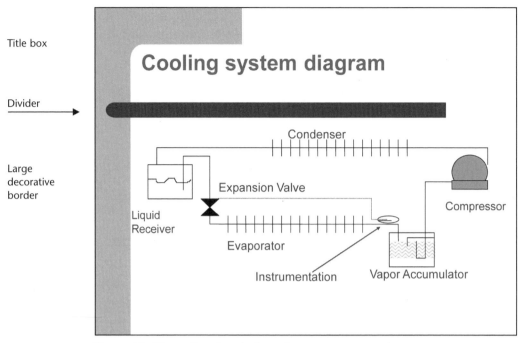

Figure 15. Heavy framing design schemes can constrain space

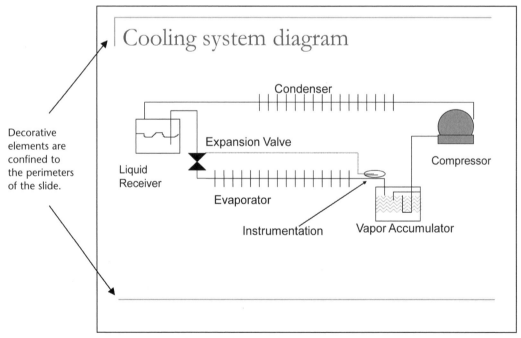

Figure 16. Light framing design schemes are usually acceptable

Chapter 3.7

Guide to Memoranda

Introduction

The memorandum is the most common type of business communication. It is the medium used by people within a company or team as they discuss business—particularly to announce policies, assign tasks, or solicit input.

Memoranda differ from letters in that they address different audiences. Letters are vehicles for external communication—with clients, for example, or regulators. For these external audiences, letters are used to describe actions that the reader should take or actions that the company may have taken in response to previous requests. Memoranda differ from reports in that reports tend to be oriented to the impersonal presentation and analysis of facts. Reports may present facts that justify action, but the actions themselves are defined and explained in the letters or memoranda to which the reports are attached.

Memos are commonly used as transmittal notes for larger reports, or they may be used to direct readers to other information sources. Memoranda may also be stand-alone documents, as when they are used to announce changes to company procedures or team task assignments.

Memoranda can be partitioned into four information sections. In order of appearance, these are the Header, the Introduction, the Details presentation, and the Action Item. These are discussed next.

1. Header

The header is the most distinctive format section of a memorandum, as it displays the well-known block of boldfaced words **To, From, Subject,** and **Date.** These should be filled out fully and specifically, as they display the lines of responsibility for the information presented in the memo, and they display the date after which the recipient can be assumed to have known his or her responsibilities with respect to the subject matter. The subject statement should be fully descriptive; it should name the work-related topic to which the memo refers, and it should name the actions that the memo requires or the changes that it will present.

2. Introduction

The text of a memorandum begins with an Introduction statement that should *name* the topic of the communication, *motivate* that topic, and state the *action* the reader will be asked to take. This is a great deal of information to present, but it must be presented very briefly. This means that your introduction should mainly name the primary concepts of your communication, leaving the details for a later section of the memo. In this way, you could say that the Introduction to your memorandum is functionally similar to an Abstract.

3. Details

The middle section of a memorandum presents details in support of—or in explanation of—the action that readers are to take. This section tends to be the longest section of the document, although its length will vary depending on the complexity of the topic and the significance of the actions that readers are asked to take. Those actions should be described briefly but specifically; readers must be explicitly notified of any steps they are expected to take, and they must have enough information to take those actions successfully. You should define actions sharply while keeping the number of those actions small, as in these action statements:

> "Please fill out the attached survey and return it to me by the end of business tomorrow."

> "Please contact Human Resources and review your W-4 as soon as possible."

In memoranda, you should avoid asking readers to perform intricate tasks. If a task involves more than one or two small actions—as when they are updating or installing a computer program—you should provide assistance, instructing readers, for example, to cooperate with a staff member who you might send to their offices to perform these tasks for them.

4. Action Items

At the close of a memorandum, requirements should be defined for all the parties to the communication. These include the actions that you are asking readers to take and the actions that you are taking in order to assure that they are successful. Statements of author and reader actions should be paired. If your readers are expected to obtain further information elsewhere—from your company's website or from a vendor, for example—you should explicitly state how they should obtain that information. Finally, at the close of any action statement, you should offer to respond to your readers' questions or to assist them in, for example, scheduling their responses, determining if they are impacted, or the like. In a workplace—even a medium-sized business—this offer should specify who in your organization will respond to individual queries and how that person is to be contacted.

You should specify a human contact for two reasons. First, the person issuing the memo may not be the person who is best able to respond to questions. Second, and related to the first, if the contact person is not specified, there is likely to be a certain amount of confusion and time lost as people with questions or concerns learn where to direct their queries.

Next we present an example memorandum that illustrates these four information sections in a way that is both concise and concrete. Each line of this document performs a function for the author and the reader, and this brief discussion will call attention to these functions.

The Heading section specifies first who is to read the memo—the staff of *S&J Mechanical*—and who is responsible for issuing the memo—the president of the company. Its date indicates when the audience is expected to have obtained the information in the memo, and the subject line describes the topic sharply. This subject line is of particular importance here, because it concretely names a topic—the company's medical benefits—as well as a significant action—*changes* in these benefits.

The Introduction paragraph presents background information in the first two sentences while naming medical benefits in particular as a topic for further discussion. The last sentence fully names the issue—changes to the company's health benefits—and an action that readers should take—to review these changes in time to take advantage of open enrollment.

The Detail section of the memo makes one significant point about each of the insurance plans mentioned in the opening section. Each of these points is accompanied by a brief evaluation, suggesting possible actions based on the change that is described.

The closing, or action section explicitly indicates what actions the company has taken to provide information to users, and it indicates what actions users can take to obtain that information. This is done by inserting the electronic address of the company's benefits page, where more detailed brochures are available for employee scrutiny. It also lists the name of a contact person, who will function as the company's expert on this matter.

CHAPTER 3.8

EXAMPLE OF A MEMORANDUM FOR WORKPLACE DISTRIBUTION

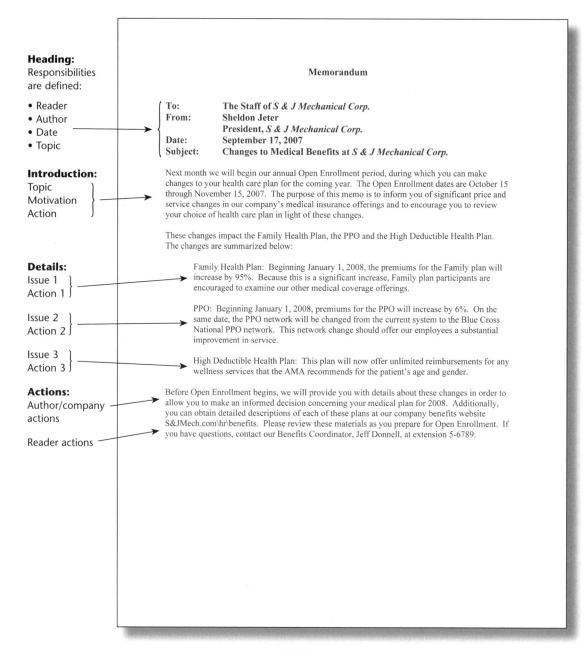

Heading:
Responsibilities
are defined:

- Reader
- Author
- Date
- Topic

Introduction:
Topic
Motivation
Action

Details:
Issue 1
Action 1

Issue 2
Action 2

Issue 3
Action 3

Actions:
Author/company
actions

Reader actions

Memorandum

To: **The Staff of *S & J Mechanical Corp.***
From: **Sheldon Jeter**
              President, *S & J Mechanical Corp.*
Date: **September 17, 2007**
Subject: **Changes to Medical Benefits at *S & J Mechanical Corp.***

Next month we will begin our annual Open Enrollment period, during which you can make changes to your health care plan for the coming year. The Open Enrollment dates are October 15 through November 15, 2007. The purpose of this memo is to inform you of significant price and service changes in our company's medical insurance offerings and to encourage you to review your choice of health care plan in light of these changes.

These changes impact the Family Health Plan, the PPO and the High Deductible Health Plan. The changes are summarized below:

Family Health Plan: Beginning January 1, 2008, the premiums for the Family plan will increase by 95%. Because this is a significant increase, Family plan participants are encouraged to examine our other medical coverage offerings.

PPO: Beginning January 1, 2008, premiums for the PPO will increase by 6%. On the same date, the PPO network will be changed from the current system to the Blue Cross National PPO network. This network change should offer our employees a substantial improvement in service.

High Deductible Health Plan: This plan will now offer unlimited reimbursements for any wellness services that the AMA recommends for the patient's age and gender.

Before Open Enrollment begins, we will provide you with details about these changes in order to allow you to make an informed decision concerning your medical plan for 2008. Additionally, you can obtain detailed descriptions of each of these plans at our company benefits website S&JMech.com\hr\benefits. Please review these materials as you prepare for Open Enrollment. If you have questions, contact our Benefits Coordinator, Jeff Donnell, at extension 5-6789.

CHAPTER 3.9

GUIDE TO E-MAIL IN THE WORKPLACE

Introduction

E-mail communications are formally similar to memoranda, with some modifications driven by the user's interface with the mail program. In the most significant modification, the memorandum heading block is replaced in e-mail messages with electronic address and subject fields; dates are automatically assigned at the time the messages are sent. When an e-mail message is viewed onscreen, these fields are commonly displayed at the head of the message, arranged as heading blocks that have a different appearance but a similar function to the heading section found in a memorandum. This memorandum layout is typically retained when e-mail messages are printed.

Like memoranda, workplace e-mail messages should be oriented toward action. Specifically, in business e-mails you should issue instructions, confirm instructions that you have received, or you should respond to questions that others have raised. You should expect your e-mail messages to become part of your professional documentation, which means that they may be preserved with your administrative files or your project records. As a result your e-mails need to characterize, for example, the issue—question or problem—you are responding to as well as your answer. Additionally, e-mails need to record the date when the issue arose and they should indicate who brought you into the discussion. This context information should be introduced—or preserved—to document the background of each of your communications.

E-mail is widely used for both private and professional purposes, and you should work carefully to keep personal e-mail separate from your professional e-mail. The best way to do this is to maintain separate e-mail accounts, using one for your business-related correspondence and the other for private communications. This helps you to avoid the serious problem of using your company's system to store or process sensitive or personal information, and it helps you to avoid embarrassment that may arise if your company's address list should become mingled with your personal address list.

Most people receive an enormous volume of e-mail. Much of this is junk that eludes the company's filters, some of it is workplace mail that has been distributed more widely than necessary, and some of it is unnecessary chatter among colleagues. To prioritize their mail, most users have developed two characteristics that greatly complicate workplace use of e-mail. The first of these is *impatience* and the second is *informality*.

1. Impatience

Users have adapted to the high volume of e-mail by learning to skim their inbox headers seeking familiar **Senders** and pertinent **Subjects.** Many people open and read messages only after a message meets their Sender/Subject criteria. They then read quickly, searching for motivation, context, and action information; readers may not finish reading notes that do not present that information in the first few lines.

As a result, your messages are screened twice—once from a mailbox-level scan of the Sender and Subject, and then again from a scan of the opening paragraph. For your mail to register on your reader, you need to make sure that these sections of your message are specific, direct, and sharply stated. As with memoranda, the subject line must specifically define the topic/action of the message. Then, in the first lines of your message, you should define why you are writing and what you want your reader to do. You should defer details about these matters to the body of your message.

2. Informality

E-mail is a common tool for informal communication. Such communication is generally conducted for fun rather than for transacting business, so the participants do not work hard to be specific, or to define actions and responsibilities. Rather, when people write informal e-mails quickly, they use nicknames and they avoid specificity in naming topics and actions. Workplace standards, however, are different from informal standards; workplace notes need to be specific, concise, and accurate. When authors accidentally compose business-related e-mails using informal standards, the result is usually unprofessional and confusing. Most informal e-mails will be ignored because they fail to present enough information for the reader.

Many of the problems related to informality can be addressed by making adjustments to your e-mail settings in order to assure that author and recipient names are fully displayed with each message. The display of recipient names can be managed by making adjustments to your electronic address book, which you can edit to display first and last names along with electronic addresses. Display of your name should be addressed in two places. First, you should assure that your first and last names appear in the "sender" column of your recipient's e-mail display without nicknames. Usually you can do this by adjusting the identity setting in your e-mailer's options or preferences menu. Additionally, you should take advantage of your mailer's ability to insert a signature at the bottom of each e-mail message. Your signature should display your full name, your job title, and your mailing address at your company, followed by your work telephone number. You should restrict your workplace signature file to official contact information; avoid aphorisms, quotations, and artwork.

While you can automate your e-mail program to manage the identity displays, your **Subject** field requires thoughtful input from you each time you compose a message. The subject line for an e-mail message performs the same job as the subject line for a memorandum; however, because readers may prioritize—or even dispose of—e-mails on the basis of the subject statement, your subject must sharply characterize both the topic of your message and the action you want the reader to take. When possible you should design subject statements that define first the project and second the action that is to be taken. The following example subject statements demonstrate how descriptive topic and action phrases can be combined to make useful subject statements:

Subject: City Hall Renovation – request for reimbursement

Subject: Atlanta Tower Crane – maintenance required

Format in E-mail

After the specifics of subject and sender have been addressed, the e-mail format resolves into a compressed version of the Memorandum, with an orienting Introduction, a detail-rich Discussion, and a brief Closing that focuses on actions the reader should take.

1. Introduction

The **Introduction** section should include a salutation line like those used in print letters, followed by an Introduction statement similar to the Introduction in a memorandum. This Introduction should compress the topic announcement and objective into a single statement of the *purpose* or *goal* of the message. If your message is prepared in response to a request from a third party—your supervisor, for example, or a client—you should introduce this as background information. The following examples suggest how an e-mail might combine a purpose statement with a request response:

> Dear Bob,
> Sheldon Jeter has asked me to respond to your query of June 23 concerning…
>
> Dear Stephen,
> I am writing in response to a request you posed to Sheldon Jeter regarding…

2. Discussion

A **Discussion** section should answer questions and explain the answers, or it should itemize the actions the reader is expected to take, offering brief explanations of those actions, as necessary. In workplace communications, all explanations must be specific. People—both colleagues in your company and contacts at other companies—should be identified by their full names; contact information such as telephone numbers, e-mail addresses, and regular addresses should be included for people who are not employees of your company. Products and purchase orders should be specified by name and identification number, and the dates and substance of previous contacts should be specified when they are cited in the message.

Readers may be asked to take a variety of actions when they read e-mail messages. They may be asked to read and respond to an attached document or electronic link, they may be asked to contact a third party, or they may be assigned a project task. Regardless of what particular action you want your reader to take, you should specify that action, you should define a date by which the reader should complete that action, and you should indicate where the reader should direct the results. The following action statements are suggestive:

> I've attached 1) a manuscript for you to review and 2) a reviewer checklist. Please respond to the questions on pages 2 and 3 of the reviewer checklist and deliver the responses to me in e-mail by Friday the 23rd.

> Please contact Bill Jones, our service representative, at 123-456-7890. Give him a purchase order number for the mass spectrometer package that we ordered and ask him to ship the device. Get a tracking number for the shipment, along with a delivery date. Get back to me before the end of the day with the shipping date and tracking number.

3. Closing

At the close of an e-mail, most authors politely offer to provide assistance to the reader. These assistance statements are commonly pleasant, but they are not always useful, as they provide phone numbers for the reader, but they do not clarify the type of assistance that the reader can obtain:

> Please let me know if I can be of further assistance. If you have questions, you can contact me at the above address, or you can call me at (123) 456-7890.

It is good to offer assistance, and such non-specific offers may be adequate for tasks that the reader is to perform for his or her own benefit. But when the reader is expected to deliver a result to a colleague, more details are usually required in the closing section. In an action-oriented e-mail you should define a specific time for completion of a task, you should define what is to be delivered when the task is completed, and you should define who receives the result or notification at the time of completion. An example closing might be this:

> I will review this using the Track Changes tool, and I'll deliver the edited file to you on Wednesday morning. Please contact me if you haven't received the file by Wednesday at noon.

Penalties for failure can be mentioned in closing sections as well:

> The deadline for submitting reimbursement requests for transactions in the 2008 fiscal year is 5:00 p.m. on June 30. Any requests for FY 2008 reimbursements that are filed after that date will not be processed.

Next we present an example e-mail exchange concerning a project task assignment. This e-mail was used to define the responsibilities of two employees and to set deadlines for completing a project and a project report.

CHAPTER 3.10

EXAMPLE OF E-MAIL EXCHANGE FOR A WORKPLACE PROJECT

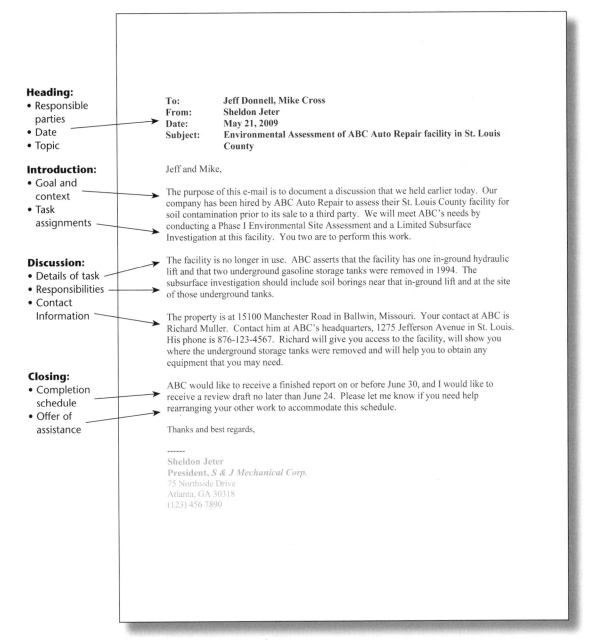

Heading:
- Responsible parties
- Date
- Topic

Introduction:
- Goal and context
- Task assignments

Discussion:
- Details of task
- Responsibilities
- Contact Information

Closing:
- Completion schedule
- Offer of assistance

To: Jeff Donnell, Mike Cross
From: Sheldon Jeter
Date: May 21, 2009
Subject: Environmental Assessment of ABC Auto Repair facility in St. Louis County

Jeff and Mike,

The purpose of this e-mail is to document a discussion that we held earlier today. Our company has been hired by ABC Auto Repair to assess their St. Louis County facility for soil contamination prior to its sale to a third party. We will meet ABC's needs by conducting a Phase I Environmental Site Assessment and a Limited Subsurface Investigation at this facility. You two are to perform this work.

The facility is no longer in use. ABC asserts that the facility has one in-ground hydraulic lift and that two underground gasoline storage tanks were removed in 1994. The subsurface investigation should include soil borings near that in-ground lift and at the site of those underground tanks.

The property is at 15100 Manchester Road in Ballwin, Missouri. Your contact at ABC is Richard Muller. Contact him at ABC's headquarters, 1275 Jefferson Avenue in St. Louis. His phone is 876-123-4567. Richard will give you access to the facility, will show you where the underground storage tanks were removed and will help you to obtain any equipment that you may need.

ABC would like to receive a finished report on or before June 30, and I would like to receive a review draft no later than June 24. Please let me know if you need help rearranging your other work to accommodate this schedule.

Thanks and best regards,

Sheldon Jeter
President, *S & J Mechanical Corp.*
75 Northside Drive
Atlanta, GA 30318
(123) 456 7890

Introduction:
Context is displayed by reproducing terms of previous message

Discussion:
• Survey schedule
• Completion date

Closing:
• Delivery dates

To: Sheldon Jeter
From: Jeff Donnell
Date: May 22, 2009
Subject: Re: Environmental Assessment of ABC Auto Repair facility in
           St. Louis County

Sheldon,

Per your request, Mike and I have contacted Richard Muller, and we have arranged to meet him at the St. Louis County facility on May 28 to begin the assessment. Based on Richard's description of the facility, the assessment should be completed that day unless more extensive borings are required. We will contact you on the afternoon of the 28th if we identify areas that require further attention from us.

We will give you a verbal update when we return on the 29th, and we will have a formal report for you by June 5.

Regards,

Jeff

Jeff Donnell
Project Manager, *S&J Mechanical Corp.*
75 Northside Drive
Atlanta, GA 30318
(123) 456 7890

Sheldon Jeter wrote:

> The purpose of this e-mail is to document a discussion that we held earlier today. Our company has been hired by ABC Auto Repair to assess their St. Louis County facility for soil contamination prior to its sale to a third party. We will meet ABC's needs by conducting a Phase I Environmental Site Assessment and a Limited Subsurface Investigation at this facility. You two are to perform this work.
>
> The facility is no longer in use. ABC asserts that the facility has one in-ground hydraulic lift and that two underground gasoline storage tanks were removed in 1994. The subsurface investigation should include soil borings near that in-ground lift and at the site of those underground tanks.
>
> The property is at 15100 Manchester Road in Ballwin, Missouri. Your contact at ABC is Richard Muller. Contact him at ABC's headquarters, 1275 Jefferson Avenue in St. Louis. His phone is 876-123-4567. Richard will give you access to the facility, will show you where the underground storage tanks were removed and will help you to obtain any equipment that you may need.
>
> ABC would like to receive a finished report on or before June 30, and I would like to receive a review draft no later than June 24. Please let me know if you need help rearranging your other work to accommodate this schedule.
>
> Thanks and best regards,
>
> ------
> Sheldon Jeter
> President, *S & J Mechanical Corp.*
> 75 Northside Drive
> Atlanta, GA 30318
> (123) 456 7890

Index